# SOFT MECHANOCHEMICAL SYNTHESIS:
## A Basis for New Chemical Technologies

# SOFT MECHANOCHEMICAL SYNTHESIS:
## A Basis for New Chemical Technologies

*by*

**Evgenii Avvakumov**
*Institute of Solid State Chemistry and Mechanochemistry*
*Siberian Branch of Russian Academy of Sciences*
*Novosibirsk. Russia*

**Mamoru Senna**
*Keio University*
*Yokohama, Japan*

**Nina Kosova**
*Institute of Solid State Chemistry and Mechanochemistry*
*Siberian Branch of Russian Academy of Sciences*
*Novosibirsk, Russia*

**KLUWER ACADEMIC PUBLISHERS**
**Boston / Dordrecht / London**

**Distributors for North, Central and South America:**
Kluwer Academic Publishers
101 Philip Drive
Assinippi Park
Norwell, Massachusetts 02061 USA
Telephone (781) 871-6600
Fax (781) 871-6528
E-Mail <kluwer@wkap.com>

**Distributors for all other countries:**
Kluwer Academic Publishers Group
Distribution Centre
Post Office Box 322
3300 AH Dordrecht, THE NETHERLANDS
Telephone 31 78 6392 392
Fax 31 78 6546 474
E-Mail <orderdept@wkap.nl>

Electronic Services <http://www.wkap.nl>

**Library of Congress Cataloging-in-Publication Data**

Avvakumov, Evgenii Grigor'evich
Soft mechanochemical synthesis : a basis for new chemical technologies / by Evgenii Avvakumov, Mamoru Senna, Nina Kosova.
p. cm.
Includes index.
ISBN 0-7923-7431-2
1. Mechanical chemistry. 2. Inorganic compounds--Synthesis. I. Senna, Mamoru II. Kosova, Nina. III. Title.

QD850 .A93 2001
541.3'9—dc21

2001034438



*Printed on acid-free paper.*

Printed in the United States of America

# Contents

# Preface

Mechanical methods of the activation of chemical processes are currently widely used for the synthesis of various compounds. The present monograph deals with the development of a novel approach to mechanochemical synthesis based on reactions of solid acids, bases, hydrated compounds, crystal hydrates, basic and acidic salts. This method has been called soft mechanochemical synthesis.

The monograph includes the papers published by the present authors. They describe the results of their investigations in the last two decades. New theoretical and experimental data on kinetics and mechanism of soft mechanochemical reactions in the mixtures of compounds mentioned above to give complex oxide compounds are presented. The description of new high energetic and high efficient mills providing effective occurrence of these reactions is delivered. The possibilities of applying soft mechanochemical synthesis for materials used in catalysis, material science, electronics, etc. are discussed. The advantages of the method proposed in comparison with other methods are demonstrated.

The monograph is designed for researchers, engineers and technicians engaged in chemical and ceramic industry, for scientists and students specialized in the area of development and application of new materials.

The experimental chapters of the monograph were equally prepared by all the authors, while the theoretical chapters mainly by Prof. E. Avvakumov.

The authors would like to thank the personnel of the Laboratory of Mechanochemical Reactions of the Institute of Solid State Chemistry and Mechanochemistry of Siberian Branch of the Russian Academy of Sciences, as well as of the Laboratory of Industrial Physical Chemistry of Keio University, for assistance in experimental investigations. They particularly thank E. Devyatkina for her help in manuscript preparation.

*E. Avvakumov*
*M. Senna*
*N. Kosova*

# Chapter 1

# INTRODUCTION TO SOFT MECHANOCHEMISTRY

The idea to perform the reactions directly between solids excluding the dissolution stage has been always attractive for chemists. However, the principle “corpora non agunt nisi fluida” (solids do not react until they have been dissolved) is still believed to be actual, therefore, usually the reactions between solids are not considered as being essential. Up to date, a major part of reaction is performed in solvents, preferably, in water. The dissolved substances, compared to the same substances in the solid state, substantially change their nature. For example, if hydrolysis occurs in the presence of water, this may result in substantial changes in the state of ions, the same as at oxidation and reduction processes. It can influence on the composition and properties of the final products.

Recently the attention to solid-phase reactions has been increased substantially due to the growing ecological requirements and the necessity to purify large amounts of wastes.

Solid-phase reactions are usually activated by high-temperature treatment [1-4]. However, the practical efficiency of this process is rather low since the diffusion rate through a product layer is small, no tight contacts between the particles of components are provided, a particle size distribution is non-uniform, etc. Therefore, the search for new methods of performance of solid-phase reactions is carried out. Some new methods are successfully used for these purposes. These methods include, for example, self-propagating high-temperature synthesis [5], shock waves [6,7], mechanical activation of mixtures in grinding apparatus [8-15]. The latter method becomes more widely used at present due to its relative simplicity and availability.

Mechanical activation of mixtures involves the dispersion of solids and their plastic deformation. These processes cause the generation of defects in solids; they also accelerate the migration of defects in the bulk, increase the number of contacts between particles, and renew the contacts. All these factors provide chemical interaction between solids initiated by mechanical loading. These factors are the subject of investigations in one of the fields of solid state chemistry, namely, the mechanochemistry of inorganic substances, which is intensively developed. In particular, a large number of reviews and monographs published on this subject is the evidence of this intensive development [8-22].

The synthesis of various compounds, including oxides, phosphates, carbides, complexes, intermetallides, alloys, has been carried out by mechanical activation. The preparation of alloys has formed a separate direction, i.e., mechanical alloying [22-28].

Oxide materials are important in various fields of technology. The methods involving mechanical activation are helpful for the preparation of these materials. However, due to high hardness of the initial anhydrous oxides, the application of mechanical activation is not very efficient yet.

It has been proposed recently to perform mechanochemical synthesis using the solids which possess higher reactivity than anhydrous oxides. These solids contain groups of atoms comprising oxygen and hydrogen. They include solid acids and bases, acidic and basic salts, crystal hydrates, i.e., the substances that react with each other releasing water. As a rule, the hardness of these compounds is 3-4 times lower than that of anhydrous oxides, that allows to decrease the level of mechanical loading and to pass to softer activation conditions. The mechanochemical reactions of hydroxides and hydrated compounds can be considered as solid-phase reactions. It can be expected that the combination of mechanical activation with the interaction according to the acid-base mechanism provides an increase of the efficiency of these reactions.

Most of the classical attempts of mechanochemical synthesis were aimed to get products directly from a mill jar. They brought to obvious drawbacks, such as contamination and energy consumption, both being very often intolerably high. In order to attract more serious interest from industry, mechanochemical processes should be inevitably shifted toward milder and more efficient conditions. One of the most rational ways is to restrict the reactions realized by mechanochemical route with unambiguous merits. At the same time, we need sound theoretical basis for mechanochemical processes to win higher credibility. When we triumph these difficulties, we may consider that mechanochemistry is capable for fabricating high value advanced materials.

Peculiarities of soft mechanochemical reactions consist in the high reactivity of surface functional groups, notably, OH groups. Most of the metal oxides absorb water molecules by their surface. Some of them are so rich with adsorbed OH groups that they are automatically utilized in industrial processes, as in the case of surface-modified $TiO_2$. However, replacing of anhydrous oxides with either hydroxides or hydrated oxides, we shall have much higher OH density per surface area. A bond energy of Me-O and O-H is a measure for determination whether an OH-group serves as an acid or a base. The acid-base reaction is realized when more than two types of Me-O(H) bonds differing by acid-base properties are brought into contact.

Addition of free water in the reaction mixture may also affect positively. However, the addition of free water cannot always be appropriate, since without sufficient chemical interaction with the substrate oxide it would not work or even can be harmful reducing the main effect of mechanical loading. Chemical interaction between solid surfaces and water under mechanical activation significantly differs from static conditions. During comminution, solids are subjected to dynamic loading that results in the extension and compression of chemical bonds. This process is believed to be similar to «corrosion under pressure» [29].

This idea was confirmed experimentally by Berschtein [29]. The solid surfaces contain potentially reactive interatomic bonds, therefore, generation and propagation of crystal fracture and under the action of stresses (applied and residual) in the presence of water proceed via mechanically induced hydrolysis. The experimental data obtained for loaded bonds in quartz in the presence or absence of water were treated according to equation

$$\tau = \tau_0 \exp ( U_0 - \gamma\sigma)/ RT , \qquad (1.1)$$

where $\tau$ is a durability of the sample, $U_0$ is an activation energy for bond rupture, $\sigma$ is a surface energy of solid, $\gamma$ is a coefficient, R is a constant.

The activation energy for Si-O bond rupture in the absence of water is 418 kJ/mol while in the presence of water 92 kJ/mol. The latter coincides with the activation energy of the hydrolysis of Si-O bond.

In solid state chemistry, a special term "soft chemistry" was proposed to describe the methods based on preliminary synthesis of intermediate compounds which are further used to prepare final products [4]. One of the authors of this monograph has introduced the term "soft mechanochemistry" for mechanochemical reactions with the participation of highly reactive compounds [30]. They include the reactions with the participation of compounds containing oxygen-hydrogen groups as well as other relatively unstable compounds such as peroxides, nitrates, metal carbonates and other reactive compounds.

Before the method has got its name, a number of papers had been published within this direction and turned out to be important for development of this method. The most important papers among them are cited below.

Boldyrev demonstrated that mechanochemical reactions of solids in the presence of water can be considered as hydrothermal ones. The products of mechanochemical synthesis for a series of reactions were found to be similar to the products of hydrothermal reactions. These reactions include the synthesis of $CaO{\cdot}FeO{\cdot}2SiO_2$ (gedenbergite) [31] and calcium hydrosilicates, eg., $5CaO{\cdot}6SiO_2{\cdot}7H_2O$ [32], as well as others.

Longo and Voght synthesized $CaMg(CO_3)_2$ (dolomite) by activation of calcium or magnesium chlorides or sulphates with $Na_2CO_3 \cdot 10H_2O$. They assumed that a rapid reaction was caused by local high solubility of components in thin films on the surfaces of ground material [33]. A similar point of view was developed also by Juhasz and Patkai-Horvat who investigated the role of water in mechanochemical reactions of silicates' formation [34 ].

Payakoff carried out the synthesis of various oxalate complexes by mechanical activation of the solid oxalic acid with hydroxides and basic salts [35]. He demonstrated that the synthesis proceeded with the formation of honey-like mass from which solid complex crystal hydrates are crystallized relatively easily under aging of activating mixtures at room temperature. Ultrafine and highly reactive catalysts, pigments and other compounds of practical importance were obtained by thermal decomposition of the mentioned complexes.

The effect of mechanical activation on the synthesis of mullite $3Al_2O_3 \cdot 2SiO_2$ from the mixtures of activated aluminium hydroxide and silica gel was studied by Klevtsov, Mastikhin, Krivoruchko et al. [36]. According to NMR, the interaction between the initial components is observed at the stage of mechanical activation. In activated mixtures, a transition of $Al^{3+}$ from octahedron into tetrahedron and pentahedron-tetrahedron ones is observed. The formation of mullite occurs under heating of activated mixtures at 1100°C, while for unactivated mixtures at 1400°C only. Lower temperature allows to obtain the product in fine disperse state ($S=50\ m^2/g$).

Fernandes-Rodriges, Morales, Tirado synthesized mechanochemically lithium ferrite and lithium cobaltite from lithium hydroxide and corresponding oxihydroxide: δ-FeOOH or CoOOH [37]. Similarly, barium hexaferrite $BaFe_{12}O_{19}$, including that doped with titanium and cobalt, was obtained from $Ba(OH)_2 \cdot H_2O$ and γ-FeOOH [38].

In the studies of mechanochemical synthesis of hydroxylapatite from the mixtures of calcium orthophosphates and oxides, Chaikina demonstrated that the synthesis rate is changed depending on the type of phosphate complex: the more protons in the complex, the higher the reaction rate. It was demonstrated that water molecules, being one of the structural components of the intermediate phase, stabilize it, that makes the rate of the formation of this phase dependent on the amount of bound water in the mixture [39 ].

The transfer from separate studies to systematic investigations and the evolution of the method of soft mechanochemical synthesis into a separate field of mechanochemistry was achieved in the works of the authors of the present monograph. It was demonstrated that mechanochemical reactions in a number of cases take place more rapidly in the mixtures of hydrated oxides than in the mixtures of anhydrous oxides [40,41]. Later on, a series of works was published to verify and develop the given method [42-46].

The goal of the present monograph is to generalize the works carried out in this research direction. The subject of investigations is the synthesis of complex oxides of the elements of I-VIII groups of the Periodic Table from ordinary hydroxides. For numerous hydroxides, substantial differences in their properties are observed, in particular, the differences in acid-base characteristics due to the structure of electron shells of atoms and the nature of chemical bonds. This allows one to search the definite laws governing the formation of complex oxides, and to look for correlation between the structure of initial hydroxides, their thermodynamic parameters and kinetics of mechanochemical synthesis.

In principle, the problem concerning the synthesis of complex oxides from the mixtures of hydrated oxides is not new. A large number of works dealt with the preparation of complex oxides by means of thermal decomposition of the mixtures of hydrated oxides including those co-precipitated and passing the stages of sol-gel process; hydrothermal treatment in aqueous medium was also used. Soft mechanochemical synthesis is distinguished from the above-mentioned methods as it affects the reaction kinetics by continuous renewal of the surface of contacts between solid particles during mechanical activation of their mixtures. Unlike heating, which is not a sufficient factor to activate solid mixtures since solid particles are spatially separated from each other, the mechanical activation makes solid particles get in contact. Such problems do not arise in solutions or gases, however, as a rule, the reactions in these media are conducted under forced mixing. Mechanical activation acts as a mixer in the reactions between solids. Thus, mechanical activation comprises the missing link in the chemistry of solid-phase reactions which did not receive sufficient attention. Mechanical activation is especially favourable for the reactions in which intermediate fluid phases are formed (solutions or melts). This comprises the similarity of soft mechanochemical synthesis with hydrothermal reactions based on the formation of soluble compounds, as well as with the reactions involving co-precipitation followed by heating. Due to the simplicity and efficiency of soft mechanochemical synthesis, the possibility of its wide application in industry is probable. A series of the examples of successful realization of this method is described in the present work.

## *References*


1. Hauffe, Karl, *Reaktionen in und Festen Stoffen*. Berlin: Springer – Verlag, 1955.
2. Schmalzried, Hermann, *Solid State Reactions*. Berlin:Verlag Chemie, 1971.
3. Tretyakov, Yurii, *Solid State Reactions*. Moskva: Khimia, 1978.
4. West, Anthony, *Solid State Chemistry and its Applications*. New York: John Wiley and Sons, 1984.
5. Mershanov A. Self-sustaining high-temperature synthesis. In *Physical Chemistry. Modern Problem*, ed. Kolotirkin Ya., Moskva: Khimia, 1983.
6. Adadurov G.A. Experimental investigations of chemical processes in conditions of dynamics pressures. Uspekhi Khimii 1986; 55: 555-70.
7. Batsanov S.S. Inorganic chemistry of high dynamic pressures. Uspekhi Khimii 1986; 55: 579-91.
8. Heinicke, Gerhard, *Tribochemistry*, Berlin: Akademie-Verlag, 1984.
9. Boldyrev, Vladimir, *Experimental Methods in Mechanochemistry of Inorganic Compounds*. Novosibirsk: Nauka, 1984.
10. Avvakumov, Evgenii, *Mechanical Methods of Activation of Chemical Processes*. Novosibirsk: Nauka, 1986.
11. Gutman, Emanuil, *Mechanochemistry of Metals and Corrosion Defence*. Moskva: Metallurgiya, 1974.
12. Tkacova, Klara, *Mechanical Activation of Minerals*. Amsterdam-Oxford-New York - Tokyo: Elsevier, 1989.
13. Kornilovich, Boris, *Structure and Surface Properties of Mechanochemically Activated Silicates and Carbonates*. Kiev: Naukova Dumka, 1994.
14. Lapteva, Ekaterina, Yusupov, Talgat, and Berger, Anna, *Physical-chemical Changes of Layered Silicates in the Process of Mechanical Activation*. Novosibirsk: Nauka. 1981.
15. Juhasz, A. and Opoczky, L., *Mechanical Activation of Minerals by Grinding: Pulverizing and Morphology of Particles*. London: Ellis Horwood, 1990.
16. Balaz, Peter, *Expactive Metallurgy of Activated Materials*. Amsterdam: Elsevier, 2000.
17. *Mechanochemical Synthesis in Inorganic Chemistry*, ed. Avvakumov E.G., Novosibirsk: Nauka, 1991.
18. Butyagin P.Y. Kinetics and nature of mechanochemical reactions. Uspekhi khimii 1971; 40: 1935-59.
19. Butyagin P.Y. Desordening structures and mechanochemical reactions in solids. Uspekhi khimii 1984; 53: 1769-81.
20. Boldyrev V.V., Avvakumov E.G. Mechanochemistry of inorganic substances. Uspekhi khimii 1971; 40: 1935-56.
21. Gilman I.I. Mechanochemistry. Science 1996; 274: 650-68.
22. Murty B.S. Ranganathan S. Novel materials synthesys by mechanical alloying/milling. Inter. Mat. Reviews 1988; 43: 101-41.
23. Fernandez-Bertran J.R. Mechanochemistry: an overview. Pure Appl. Chem. 1999; 71: 581-86.
24. Gaffet E., Tillement O. Mechanochemistry and mechanical activation. Annales de Chimie 1997; 22: 417-37.
25. Benjamin J.S. Mechanical alloying. Scientific American 1976; №5: 40-8.
26. Oleszak D. Mechanical alloying. A novel method for synthesis and processing of materials. Acta Phys. Pol. A 1999; 96: 101-12.
27. Koch C. Intermetallic matrix composites prepared by mechanical alloying. A review. Mater. Sci. Eng. A 1998; 224: 39-48.
28. El-Eskandarani, Sherif, *Mechanical Alloyng: Fabrication of Advanced Materials at Room Temperature*. Cairo-Egipt: DAR Al-FIKR AL-ARABI, 1999.
29. Berschtein, Vladimir, *Mechanohydrolitic Processes and Strength of Solids*. Leningrad: Nauka, 1987.

30. Senna M. Incipient chemical interaction between fine particles under mechanical stress - a fearsibility of producing advanced materials via mechanochemical routes. Solid State Ionics 1993; 63-65: 3-9.
31. Boldyrev V.V. About kinetic factors determining the specific of mechanochemical reactions in inorganic systems. Kinetika i Kataliz 1972; 13: 1411-21.
32. Boldyrev V.V., Khabibullin A. Kh., Kosova N.V., Avvakumov E. G. Hydrothermal reactions under mechanochemical treatment. J. Mater. Synthesis and Processing 1996; 4: 377-81.
33. Longo J.M., Voight K.C. Synthesis of mixed metal carbonates by grinding. Solid State Ionics 1989; 32/33: 409-13.
34. Juhasz A.Z., Patkai- Khorvat M. Mechanochemical reactions in magnesia-silikon-water system. Hung. J. Industr. Chem. 1988; 16: 439-47.
35. Payakoff Sw. Mechanochemicshe Reaktionen. Osterr. Chemie Zeit. 1985; 3: 48-56.
36. Klevtsov D.P., Zolotovskii B.P., Krivoruchko O.P., Buyanov R.A. Interaction in alumosilicate mixtures during mechanical and thermal treatment. Zhurn. Prikladnoi Khimii 1988; 61: 913-14.
37. Fernandes-Rodrigues J.M., Morales J., Tirado J.L. Mechanochemical preparation and degradation of $LiCoO_2$. Reactivity of Solids 1987; 4: 163-72.
38. Shapkin V.L., Kormilitsina Z.A., Avvakumov E.G. Mechanochemical synthesis of amorphous and crystalline barium ferrites. Proc. Conf. «Mechanochemical synthesis». Vladivostok: DVGU , 1990.
39. Chaikina M.V. Mechanochemical synthesis of apatites and ortophosphates. In *Mechanochemical Synthesis in Inorganic Chemistry*, ed. Avvakumov E.G. Novosibirsk: Nauka, 1991.
40. Avvakumov E.G., Kosova N.V. Fast propagating solid-state mechanochemical reactions. Sib. Khim. Zhurn. 1991; 5: 62-6.
41. Avvakumov E.G., Kosova N.V., Devyatkina E.T. Mechanochemical reactions of hydrated oxides. J. Solid State Chem. 1994; 113: 379-83.
42. Avvakumov E.G., Kosova N.V. Soft mechanochemical synthesis: specific features and outlook. In *Chemistry Reviews. Advances in Mechanochemistry: Physical and Chemical Processes under Deformation*, ed. Butaygin P.Yu., Dubinskaya A.M., Amsterdam: Harwood Academic Publisher, 1998 ; 23: 285-312.
43. Watanabe T., Isobe T., Senna M. Mechanism of incipient chemical reaction between $Ca(OH)_2$ and $SiO_2$ under moderate mechanical stressing. A solid state acid-base reactions and charge transfer due to complex formation. J. Solid State Chem. 1996; 122: 74-9.
44. Watanabe T., Isobe T., Senna M. Examination of radical mechanism by an EPR study. J. Solid State Chem. 1996; 122: 291-96.
45. Watanabe T., Isobe T., Senna M. Changes in short range ordering throughout the mechanical and thermal processes. J. Solid State Chem. 1997; 130: 284-89.
46. Temuujin J. Okada K. Mackenzie K.Y.D. Role of water in the mechanochemical reactions of $MgO$-$SiO_2$ systems. J. Solid State Chem.1998; 138: 169-77.

# Chapter 2

# CHEMICAL BONDS, STRUCTURE, PROPERTIES AND PREPARATION OF HYDRATED OXIDES

## 2.1. Chemical bonds in hydrated oxides and their structures

The behavior and properties of hydrated oxides are determined by fundamental characteristics of elements:

a) electron structure of atom and the composition of outer electron shell of ion formed from this atom, which are determined, in turn, by the position of the element in Mendeleev's Periodic Table;

b) ion charge and radius, that define charge density of the cation;

c) ionization potential, affinity to electron and electronegativity of the atom.

For the compounds considered in the present work, electron configurations of the outer electron shells that participate in the formation of chemical bonds are as follows:

| | | |
|---|---|---|
| Li | $1s^22s^1$ | |
| Mg | $1s^22s^22p^23s^2$ | |
| Al | | $3s^23p^1$ |
| Si | | $3s^23p^2$ |
| P | | $3s^23p^3$ |
| Ca | $1s^22s^22p^63s^23p^6$ | $4s^2$ |
| Ti | | $3d^24s^2$ |
| V | | $3d^34s^2$ |
| Mn | | $3d^54s^2$ |
| Fe | | $3d^64s^2$ |
| Co | | $3d^74s^2$ |
| Ni | | $3d^84s^2$ |
| Cu | | $3d^{10}4s^2$ |
| Zn | | $3d^{10}4s^2$ |
| Zr | $1s^22s^22p^63s^23p^63d^{10}4s^24p^6$ | $4d^25s^2$ |
| La | | $4d^{10}5s^25d^16s^2$ |
| W | | $4d^{10}5s^25d^46s^2$ |
| Pb | | $4d^{10}4f^{14}5s^25p^65d^{10}6s^26p^2$ |

Among the listed elements, Li, Mg and Ca are s-elements, Al, Si, P are p-elements, Ti, V, Mn, Fe, Co, Ni, Cu, Zn, Zr, La, W are d-elements, since they contain the mentioned electrons in their outer electron shell.

The formation of chemical bond between atoms is accompanied by the rearrangement of electron shells of atoms; atomic orbitals give rise to molecular orbitals in molecules and to energy zones, similar to molecular orbitals, in crystals.

The main factors that determine the steric coordination of molecules and crystal lattices (accepting that the positions of chemical bonds between atoms in lattice obey the same laws of electron density distribution that are actual for molecules) is the direction and the principle of maximum overlapping. The strongest chemical bonds are formed in the direction of maximum overlapping between the orbitals of binding electrons.

Valence electrons are characterized by definite shapes of electron orbitals. Hence, atoms with s-type valence electrons can form the bonds of similar strength in any direction, while all the directions being equivalent. For p-electrons, the strongest overlapping occurs in the direction of the dumbbell axis. So, the angle between two bonds formed by p-electrons of one atom with s-, p-, d-electrons of two other atoms should be 90°. For d-electrons, the position of bonds is square or octahedral.

If valence electrons of the atoms forming chemical bond are in different states, one could expect the formation of different types of bonds. However, experiments show that equivalent bonds are formed. In such cases, the state of electrons is described not by pure s-, p-, d- or f-functions but by mixed hybrid wave functions that are linear combinations of basic functions describing the state of electrons. The combination of two electrons in s- and p-states brings to the formation of hybrid bonds, the angle between them being 180°C; one s- and two p-electrons give rise to three hybrid bonds, angles between them being 120°C; one s- and three p-electrons give four bonds at an angle of 109°28. Atoms with d- and f-electrons form more complicated configurations.

When crystals are formed from atoms, discrete energy levels are substituted by the bands incorporating large number of discrete levels of 1s, 2s and other electrons. The outer electrons of a shell form conduction band while the inner electrons form valence band. Depending on the width of band gap (the distance between valence band and conduction band), a solid exhibits either conducting or dielectric properties.

The direction of a chemical bond directly determines the structure of covalent compounds. For example, in diamond (its electron configuration being $1s^2 2s^2 2p^2$), four hybrid $sp^3$ orbitals are formed due to the destruction of a spin bond at s levels and the excitation of three electrons at p levels, they are directed from the centre to the vertices of regular tetrahedron. The angle between the axes of orbitals is equal to 109°28 .

For ion crystals, the structure is determined by the directions of the outer filled electron shells of ions. Sodium chloride can be considered as an example. Electron configurations of free sodium and chlorine atoms are as follows: $1s^2 2s^2 2p^6 3s^1$ (sodium) and $1s^2 2s^2 2p^6 3s^2 p^5$ (chlorine). The formation of NaCl compound proceeds by transferring a weakly bound 3s electron of a sodium atom to 3p-orbital of a chlorine atom, which results in the formation of a positively charged $Na^+$ ion and negatively charged $Cl^-$ ion, both possessing completely filled outer $p^6$ orbitals. These ions come closer to each other due to the electrostatic attraction, and their outer p-orbitals shaped as orthogonal dumb-bells interact with each other through their outer ends. This overlapping and exchange interaction due to orthogonality of p-orbitals brings to the cubic cell structure.

There is no sharp difference between covalent and ion types of bonds. In covalent compounds, electron density is spread almost symmetrically over the partners, its mass centre being in the middle of the distance between atoms, while in purely ionic compounds the maximum of binding electron orbital coincides with the centre of one of the atoms. Polar bonds exhibit asymmetry in the distribution of electron density, mass centre being shifted from the middle to one of the partners. This shift is characterized by the degree of the bond's ionicity.

It was found possible to characterize the degree of the bond's ionicity by atomic electronegativity.

Pauling was the first to introduce the concept of electronegativity (EN). He defined it as a force by which an atom in a molecule attracts electrons [1]. To introduce this characteristic, thermochemical data on the energies of molecule formation ($\Delta H_{form}$) were used to calculate average bond energies. For example, for $AB_n$ molecule (where A is cation and B is anion), average bond energy is

$$E(A - B) = \Delta H_{form}(AB_n)/n \quad (2.1)$$

Pauling assumed that the energy of covalent A—B bond would be average of the energies of the corresponding uniform bonds

$$E_{cov}(AB) = 1/2[E(AA) + E(BB)] \quad (2.2)$$

In the case of polar bond, its energy will be more than the covalent (additive) value by the ionic term $\Delta E(AB)$

$$\Delta E(AB) = E(AB) - 1/2\ [E(AA) + E(BB)]. \quad (2.3)$$

This value is practically equal to the heat effect of the reaction

$$1/2A_2 + 1/2B_2 = AB + Q \quad (2.4)$$

Pauling discovered that the function $\Delta E^{1/2}$ (or $Q^{1/2}$) is atomically additive and it can be considered as the characteristic of the electronegativity of the atoms that participate in the formation of a bond, according to the equation:

$$X_A - X_B = 0.208\Delta E^{1/2}\ (\text{eV}), \qquad (2.5)$$

where $X_A$ and $X_B$ are electronegativities of A and B atoms. Having accepted the electronegativity of hydrogen to be $X_H = 2.1$, Pauling calculated X for a large number of elements, which formed the basis of the first scale of electronegativity. This equation is valid only if $\Delta E(AB)>0$. In order to overcome this shortcoming, Pauling proposed to replace $S[E(AA) + E(BB)]$ by $[E(AA)\cdot E(BB)]^{1/2}$:

$$\Delta E(AB) = E(AB) - [E(AA)\cdot E(BB)]^{1/2} \qquad (2.6)$$

As a result of this replacement, $\Delta E(AB)$ is always positive, while the results are practically identical to those obtained using the primary equation. Pauling's scale has become generally acknowledged. It is believed to be not only the most suitable one but also sufficient. Pauling's method of electronegativity determination is one of the basic ones till present time.

Thermochemical calculation of EN is not a single method to determine energy characteristics of atoms. Calculations of electronegativity using the potentials of ionization and affinity to electron are also widely used. Ionization potentials are important experimental quantitative characteristics of the stability of electron configurations of separate atoms and ions formed from them. Ionization potential characterizes the strength of electron binding to atomic core being equal to the work against their interaction forces:

$$I_n = A_n/e \qquad (2.7)$$

where e is an electron charge, n means the first, second, etc. ionization potential of the atom.

The ionization of elements is followed by a substantial increase of $I_n$ when passing from p-elements to s-elements (2–4 eV for p-elements and 10–30 eV for s-elements). On contrary, ionization potentials of p- and d-elements are close to each other (0.3–0.2 eV for d-elements and 2–4 eV for p-elements).

On the other hand, when an electron is joined to an atom, energy equal to the energy of electron binding to atom is released. It is called "electron affinity energy" ($F_n$) and is determined experimentally as the energy consumed for removing an electron from negatively charged ion, i.e.

as ionization potential of the anion. Usual range constitutes (-0.2) ÷(+3.6) eV.

Using the method of valence bonds, Mulliken showed that the electronegativity (per one electron) can be expressed by the equation [2]

$$X = 1/2(I_n + F_n). \quad (2.8)$$

This was the basis for the "absolute" electronegativity scale which was linearly linked with Pauling's scale by the equation:

$$X_{Pauling} = 0.336\,(X_{abs} - 0.615) \quad (2.9)$$

The definition of electronegativity, according to Mulliken, assumes its dependence on the valence state of atoms that allows one to reject the idea of a single EN for each atom independent of its hybridization. Thus, the necessity arises to take into account an orbital in which the valence electron is localized.

When investigating this question, Hinze and Jaffe [3] introduced the concept of "orbital" electronegativity and developed a semi-empirical method to calculate the ionization potential ($I_v$) and electron affinity ($F_v$) of any valence state. This allowed to determine electronegativity according to Mulliken's procedure taking into account the corresponding valence states of atoms

$$X = (I_v + F_v)/2 \quad (2.10)$$

It should be noted that quantum chemical calculations are used to determine electronegativity of pure s-, p- and d-states. Then, to obtain the atom's electronegativity, they are summarized according to the degree of atom orbital hybridization

$$X = n_sX_s + n_pX_p + n_dX_d \quad (2.11)$$

For example, for $sp^3$ hybridization,

$$X_{sp3} = 1/4(X_s + 3X_p) \quad (2.12)$$

The idea of orbital EN was further developed by Iczkowski and Margrave [4]. They represented the energy of an isolated atom (E) as a function of electron shell filling:

$$E(s) = E_o + a\delta + b\delta^2 + c\delta^3 + d\delta^4, \quad (2.13)$$

where $E_o$ is the energy of the ionized core, $\delta$ is a partial charge of the atom, a, b, c, d are empirical coefficients depending on the nature of the atom and on its valence state.

According to the authors opinion, it is sufficient to limit the consideration to the terms of the second order in the equation 2.13 (at least for s- and p-elements):

$$E(s) = E_o + a\delta + b\delta^2 \quad (2.14)$$

Iczkowski and Margrave considered E(s) as a continuous and differential function of $\delta$; they also identified the EN of an isolated atom as the first derivative of energy with respect to the number of electrons filling the valence orbitals

$$X = dE/d\delta = a + 2b\delta \quad (2.15)$$

This means that electronegativity is a linear function of $\delta$. For neutral atoms ($\delta = 0$) X is equal to a. For charged atoms (ions), the effect of charge on the atom electronegativity is more significant due to the term $b\delta$.

The main difficulty in the described method of EN determination consists in the calculation of Iv and Ev values that cannot be determined experimentally. The description of their calculation procedures comes out of the frames of the present consideration.

Therefore, in order to solve the questions concerning the degree of ionic character of a bond, its strength, etc., one uses the values of Pauling's electronegativity or absolute electronegativities which are shown in Table 2.1 (the values are taken from the Reference Book [5]).

*Table 2.1.* Absolute electronegativities of elements [5].

| Periods of Mendeleev's Periodic Table | | | | | | | | | | | |
|---|---|---|---|---|---|---|---|---|---|---|---|
| 1 | | 2 | | 3 | | 4 | | 5 | | 6 | |
| H | 7,18 | Li | 3.01 | Na | 2.85 | K | 2.42 | Rb | 2.34 | Cs | 2.18 |
| | | Be | 4.90 | Mg | 3.75 | Ca | 2.20 | Sr | 2.00 | Ba | 2.40 |
| | | | | | | Sc | 3.34 | Y | 3.19 | La | 3.10 |
| | | | | | | Ti | 3.45 | Zr | 3.64 | Hf | 3.80 |
| | | | | | | V | 3.60 | Nb | 4.00 | Ta | 4.11 |
| | | | | | | Cr | 3.72 | Mo | 3.90 | W | 4.40 |
| | | | | | | Mn | 3.72 | Te | 3.91 | Re | 4.02 |
| | | | | | | Fe | 4.06 | Ru | 4.50 | Os | 4.90 |

| | | | | | |
|---|---|---|---|---|---|
| | | | Co 4.30 | Rh 4.30 | Ir 5.40 |
| | | | Ni 4.40 | Rd 4.45 | Pt 5.60 |
| | | | Cu 4.48 | Ag 4.44 | Au 5.77 |
| | | | Zn 4.45 | Cd 4.33 | Hg 4.91 |
| | B 4.29 | Al 3.23 | Ga 3.20 | In 3.10 | Tl 3.20 |
| | C 6.27 | Si 4.77 | Ge 4.60 | Sn 4.30 | Pb 3.90 |
| | N 7.23 | P 5.62 | As 5.30 | Sb 4.85 | Bi 4.69 |
| | O 7.54 | S 6.22 | Se 5.89 | Te 5.49 | Po 5.16 |
| | F 10.41 | Cl 8.30 | Br 7.59 | I 6.76 | At 6.20 |
| He 12.30 | Ne 10.60 | Ar 7.70 | Kr 6.79 | Xe 5.84 | Rn 5.10 |

On the basis of electronegativity concept, Pauling formulated the following rule: the larger is the difference between the electronegativities of elements, the higher is the energy of the bond between atoms.

The degree of bond ionicity in compounds is estimated using the equation proposed by Pauling:

$$i = 1 - e^{-0.18}{}_{\Delta}{}^{X} \qquad (2.16)$$

where i is the ionicity degree, $\Delta X = X_A - X_B$, where $X_A$, $X_B$ are electronegativities (according to Pauling) of the atoms forming a bond.

Recent years, approaches for the calculations of the electronegativities of the entire groups of atoms, radicals and molecules have been developed. According to oxides, Viting introduced a concept of average orbital electronegativity of a compound, which is estimated according to the additivity principle [6]. I.e., average orbital electronegativity of a compound is equal to:

$$X = \Sigma n_i X_i / \Sigma n_i, \qquad (2.17)$$

where $n_i$ is a number of atoms in a compound, $X_i$ is an absolute electronegativity of an atom (according to Mulliken).

Average orbital electronegativity is based on the Sanderson's postulate of energy levelling and averaging of contributions from atoms which form a molecule [7]. This energy levelling principle is widely used in quantum mechanics to describe orbital hybridization during the formation of chemical bonds [8].

Table 2.2 shows the values of average orbital electronegativities for some oxides and hydroxides, calculated according to the equation 2.17. One

*Table 2.2.* Average orbital electronegativities of some oxides and hydroxides calculated according to the equation 2.17.

| Absolute electronegativity | | Average orbital electronegativity of oxide | | Average orbital electronegativity of hydroxide | |
|---|---|---|---|---|---|
| atom | $X_{atom}$ | oxide | $X_{oxide}$ | hydroxide | $X_{hydroxide}$ |
| Cs | 2.18 | $Cs_2O$ | 3.95 | CsOH | 5.63 |
| Rb | 2.34 | $Rb_2O$ | 4.07 | RbOH | 5.64 |
| K | 2.42 | $K_2O$ | 4.13 | KOH | 5.67 |
| Na | 2.85 | $Na_2O$ | 4.41 | NaOH | 5.82 |
| Li | 3.02 | $Li_2O$ | 4.52 | LiOH | 5.90 |
| Sr | 2.00 | SrO | 4.71 | $Sr(OH)_2$ | 6.03 |
| Ca | 2.20 | CaO | 4,87 | $Ca(OH)_2$ | 6.30 |
| Ba | 2.40 | BaO | 4.97 | $Ba(OH)_2$ | 6.35 |
| Mn | 3.72 | MnO | 5.63 | $Mn(OH)_2$ | 6.55 |
| Mg | 3.75 | MgO | 5.64 | $Mg(OH)_2$ | 6.64 |
| Fe | 4.24 | FeO | 5.89 | $Fe(OH)_2$ | 6.73 |
| Mn | 3.72 | $Mn_2O_3$ | 6.02 | $Mn(OH)_3$ | 6.73 |
| La | 3.10 | $La_2O_3$ | 5.76 | $La(OH)_3$ | 6.74 |
| Ga | 3.20 | $Ga_2O_3$ | 5.81 | $Ga(OH)_3$ | 6.76 |
| Al | 3.23 | $Al_2O_3$ | 5.82 | $Al(OH)_3$ | 6.77 |
| Be | 4.90 | BeO | 6.07 | $Be(OH)_2$ | 6.80 |
| Cr | 3.72 | $Cr_2O_3$ | 6.01 | $Cr(OH)_3$ | 6.85 |
| Fe | 4.24 | $Fe_2O_3$ | 6.22 | $Fe(OH)_3$ | 6.91 |
| Zr | 3.64 | $ZrO_2$ | 6.24 | $Zr(OH)_4$ | 6.94 |
| Ti | 3.45 | $TiO_2$ | 6.18 | $Ti(OH)_4$ | 6.94 |
| Ge | 4.60 | $GeO_2$ | 6.56 | $Ge(OH)_4$ | 7.08 |
| Si | 4.77 | $SiO_2$ | 6.62 | $Si(OH)_4$ | 7.06 |
| V | 3.60 | $V_2O_5$ | 6.41 | $HVO_3$ | 6.70 |
| W | 4.40 | $WO_3$ | 6.75 | $H_2WO_4$ | 6.96 |
| P | 5.62 | $P_2O_5$ | 6.99 | $H_3PO_4$ | 7.08 |
| S | 6.22 | $SO_3$ | 7.21 | $H_2SO_4$ | 7.25 |
| N | 7.23 | $N_2O_5$ | 7.44 | $HNO_3$ | 7.40 |

can see that they change more regularly comparing to atomic electronegativities.

Though the concept of orbital EN allowed to unite many general concepts, it still contains many questions. In particular, the levelling principle is discussed, as well as specific levelling algorithm (arithmetic or geometric average), etc.

A more exact solution of the problems dealing with chemical bonds has been achieved in the frames of the density functional theory which has been proposed recently. This theory is dynamically developing nowadays [9]. Without considering the details of this theory, we shall only state that it involves the electron chemical potential which is the same for the entire system and is defined as a partial derivative of the electron energy of the system E with respect to the number of electrons N at a constant external potential V. It was assumed [10] that the negative electron chemical potential is identical to electronegativity. The derivative of chemical potential with respect to the number of electrons in the system (the second derivative of the energy of the system) was defined as the absolute rigidity of the system. Having assumed a square dependence of the system energy on the number of electrons, the authors [11] expressed the electronegativity of X atom using the ionization potential and electron affinity F as

$$X = (I + F)/2 \qquad (2.18)$$

and their half-difference as the hardness of the system

$$\eta = (I - F)/2. \qquad (2.19)$$

Thus, the definition of EN in the frames of density functional theory coincides with Mulliken's idea of absolute EN. Besides, a new parameter $\eta$ is introduced. It is defined as the absolute rigidity [11].

Studying acid-base properties of acids and bases, Pearson put forward the principle of "hard"and "soft" acids and bases according to which hard acids prefer to react with hard bases, while soft acids prefer to react with soft bases [12,13]. This principle has been widely used in chemistry, however, no quantitative characteristics of the hardness of acids and bases were known.

The definition of the hardness $\eta$ according to the density functional theory easily demonstrates the essence of this parameter. According to Klopman [8], the potential I and electron affinity F can be expressed through the energies of boundary orbitals: upper occupied (T) and lower vacant (L) ones

$$E_T = I \quad \text{and} \quad -E_L = F. \qquad (2.20)$$

The hardness of the system according to the equation (2.19) is determined by half-difference between energy levels

$$\eta = (E_L - E_T)/2\,. \qquad (2.21)$$

Hence, a "hard" molecule should possess a wide energy gap between the lower and upper orbitals, while in a "soft" molecule this gap should be narrow. This explains the difference in their polarizability.

Another important postulate, put forward by Pearson [13], is the principle of maximum hardness, according to which a system tends to attain the maximum rigidity. This principle was based also on experimental observations. According to (10), $\eta$ increases with increasing of ionization potential and with decreasing of electron affinity. Thus, the system tends neither to render its own electrons, no to get foreign ones, i.e. to remain stable.

Hydroxides $M(OH)_n$ comprise a numerous class of compounds ranging from strongly basic hydroxides of alkaline metals and alkaline earths, to the so-called amphoteric hydroxides (of beryllium, aluminium, zinc and others) and the hydroxides of transition metals, and further to hydroxo-acids formed by non-metals or semi-metals.

Crystal structures of hydroxides and hydrated oxides are in close connection with the nature of chemical bonds in them. Hydroxides formed by the cations with small charge and low polarizing ability possess ion lattice. They include sodium and potassium hydroxides. With increasing charge and decreasing of a cation radius, layered structures appear (LiOH, $Mg(OH)_2$, $Ca(OH)_2$, etc.). They are characterized by three-layered packings. The $OH^-$ anions compose the first and the third layer, being laid one on another, according to the principle of the closest hexagonal packing. The cations occupy the second layer and are located in the octahedral sites formed by anions (three out of each layer).

For the compounds with amphoteric properties, the existence of coordination groups $Me(OH)_4$ (tetrahedrons) and $Me(OH)_6$ (octahedrons) is characteristic; hydroxyl bonds exist between separate groups. Molecular structures are characteristic of the compounds with clearly expressed acidic properties. This means that lattice sites are occupied by molecules bound to each other by hydrogen bonds. Since this leads to the decrease of lattice strength, some acids are liquids under normal conditions, while all the bases and amphoteric hydroxides are solids.

Hydrogen bond arises as a result of the hydroxyl ion behaviour in the vicinity of a cation. Three types of $OH^-$ ion behaviour are possible: a) as an ion with efficient spherical symmetry; b) with cylindrical symmetry; c) polarized in a manner causing tetrahedral charge distribution [14].

The formation of hydrogen bonds occurs in the case c. The transition from (b) to (c) is observed with increasing of cation charge. Polarizing ability of $M^{n+}$ cation is proportional to $ne/r^2$. For $Ca^{2+}$, $La^{3+}$, $Al^{3+}$ it is 1.8, 2.0, 9.1, respectively, i.e. for $Al^{3+}$ it is much higher than for $Ca^{2+}$ and $La^{3+}$. Hence, it is clear why hydrogen bonds exist in $Al(OH)_3$ but not in $Ca(OH)_2$ or $La(OH)_3$.

Structural characteristics for a series of hydroxides are shown in Table 2.3 [15].

*Table 2.3.* Structural characteristics of hydroxides and hydrated oxides [15].

| Hydroxide | Syngony | Lattice parameters, nm | | | | Space group | Number of coord-inated $OH^-$ groups |
|---|---|---|---|---|---|---|---|
| | | *a* | *b* | *c* | β, ° | | |
| LiOH | tetragonal | 0,355 | | 0,435 | | P4/nmn | 4 |
| LiOH $H_2O$ | monoclinic | 0.737 | 0.826 | 0.319 | 110° | 2/m | |
| $Mg(OH)_2$ | hexagonal | 0.315 | | 0.476 | | P$\bar{3}$m1 | |
| $Ca(OH)_2$ | hexagonal | 0.358 | | 0.489 | | P$\bar{3}$m1 | 6 |
| $Mn(OH)_2$ | hexagonal | 0.334 | | 0.468 | | - | 6 |
| $Fe(OH)_2$ | hexagonal | 0.324 | | 0.447 | | P$\bar{3}$m1 | 6 |
| α-FeO(OH) | rhombic | 0.460 | 0.994 | 0.300 | | Pbnm | |
| α-$Co(OH)_2$ | hexagonal | 0.309 | | 0.800 | | P$\bar{3}$m1 | |
| β-$Co(OH)_2$ | hexagonal | 0.317 | | 0.464 | | P$\bar{3}$m1 | |
| CoOOH | trigonal | 0.285 | | 1.315 | | R$\bar{3}$m | |
| $Ni(OH)_2$ | hexagonal | 0.312 | | 0.450 | | C3m | |
| β-NiO(OH) | hexagonal | 0.304 | | 0.452 | | C3m | |
| $Cu(OH)_2$ | rhombic | 0.295 | 1.058 | 0.526 | | Cmcm | 6 |
| Σ-$Zn(OH)_2$ | rhombic | 0.517 | 0.855 | 0.493 | | $P2_12_12_1$ | 4 |
| $Al(OH)_3$ γ - gibbsite | monoclinic | 0.864 | 0.507 | 0.472 | 94.34 | $P2_1/n$ | 6 |
| $Al(OH)_3$ $γ^1$-gibbsite | triclinic | 1.733 | 1.008 | 0.373 | 94.17 | | 6 |
| $Al(OH)_3$ bayerite | monoclinic | 0.501 | 0.868 | 0.476 | 90.27 | $P2_1/n$ | 6 |

| | | | | | | | |
|---|---|---|---|---|---|---|---|
| $Al(OH)_3$ norstrandite | triclinic | 0.889 | 0.500 | 1.020 | 100.23 | | 6 |
| AlOOH boemite | ortho-rhombic | 0.297 | 1.223 | 0.570 | | P2/m2/m2/m | 5 |
| AlOOH diaspare | ortho-rhombic | 0.440 | 0.943 | 0.284 | | Pbnm | 5 |
| $La(OH)_3$ | hexagonal | 6.523 | 3.855 | | | $P6_3/m$ | 9 |
| $SiO_2 \cdot nH_2O$ | X-ray amorphous | | | | | | |
| $TiO_2 \cdot nH_2O$ | X-ray amorphous | | | | | - | |
| $H_2V_2O$ | ortho-rhombic | 0.356 | 0.437 | 1.155 | | - | |
| $H_2WO_3$ | ortho-rhombic | 0.514 | 1.070 | 0.512 | | Pmnb | |
| $H_3PO_4 \cdot 0{,}5H_2O$ | monoclinic | 0,792 | 1,299 | 0.747 | 109.9 | - | |

It is not quite correct to write chemical formulas as $MeO \cdot H_2O$, $Me_2O_3 \cdot H_2O$ instead of $Me(OH)_2$, $Me(OH)_3$, since structural data suggest that hydroxides are composed of $Me^{2+}$ or $Me^{3+}$ cations, and $OH^-$ groups, while oxyhydroxides are composed of $OH^-$, $O^{2-}$ and $Me^{n+}$ ions. Nevertheless, in order to simplify the presentation, we use this form of writing in some cases.

## 2.2. Properties of hydrated oxides

### 2.2.1. Acid-base properties

Different approaches are used to explain acid-base properties of substances. According to Brönsted and Lowry [16,17], an acid is a substance able to act as proton donor, while a base is a substance able to act as proton acceptor. According to Lewis [18], acids are the compounds that are able to accept electron pairs forming covalent bond, while bases are the compounds that act as donors of electron pair.

Acid-base properties of cations incorporated into the compounds are dependent to a great extent on the cation charge density. This is explained by the fact that in the compounds containing oxygen-hydrogen groups, the cations of large radius and low valency, i.e. with low electron density, are bound to oxygen weaker than to hydrogen. Therefore, the dissociation of hydroxides, for example, occurs according to the scheme $MeOH \rightarrow Me^+ +$

cations form another limiting group. In these compounds the cation is bound to oxygen much stronger than to hydrogen. Therefore, they dissociate releasing $H^+$, according to the scheme $HMeO_3 \rightarrow H^+ + MeO_3^-$, thus, being related to acids. Cations with amphoteric properties occupy an intermediate position between the two limiting cases. It was proposed to characterize the compounds quantitatively using ion potential F which is a ratio of the charge of central ion to its radius $Z^{n+}/r$, where $Z^{n+}$ is a charge of a cation and r is its radius. If $\sqrt{F} < 2.2$, a hydroxide is considered to be a base; for $2.2 < \sqrt{F} < 3.2$, a hydroxide is amphoteric; for $\sqrt{\Phi} > 3.2$, it is an acid [19].

The strength of bases and acids in solution is characterized by their dissociation constants

$$K_a = [H^+][MeO_3^-]/[HMeO_3]; \quad K_b = [M^+][OH^-]/[MOH]. \quad (2.22)$$

Their acidic or basic properties are characterized by $pH = -lg[H^+]$.

For solid acids and bases, the acidic and basic properties can also be expressed by a similar equation, but it incorporates the concentrations of acidic or basic sites present on the surface instead of concentrations of hydrogen or hydroxyl ions:

$$- lg[H^+_{surf.}] = - lgKa - lg[MO_3^-]/[HMeO_3] = H_0$$
$$- lg[OH_{surf.}] = - lgKb - lg[M^+]/[MOH] = H_- \quad (2.23)$$

$H_0$ and $H_-$ are called Hammett functions [20].

The acidic and basic properties of aqueous solutions are changed within the pH range from 0 to 14, while for solid acids and bases, they can be changed within the range $-30 < H_0, H_- < +40$. A boundary point between $H_0$ and $H_-$ is $H_0 = 7$. One can see that these limits are much broader than in the case of aqueous solutions. Acids with $H_0$ value within $7 \div -12$ correspond to usual acids, while above $-12$, they are superacids. Similarly, the bases with $H_-$ from 7 to 26 correspond to usual bases while above 26 they are superbases. The concept of superacids and superbases is widely used to explain the processes that take place in acid-base catalysis [21].

The strength of solid acids and bases is characterized by proton affinity (PA). For a base B, PA is equal to the enthalpy of reaction $B + H^+ \rightarrow BH^+$ in gas phase, where B is electrically neutral base and $BH^+$ is its protonated form. The methods have been developed to determine PA for various compounds by the combined application of different indicators, sorbents and IR spectroscopy methods [22].

The monograph [22] presents the results on PA determination for some oxides using IR spectroscopy data (according to the positions of bands of OH groups on oxide surfaces). They are equal to (in kJ/mol): MgO - 1560, CaO - 1870 and 1740 (1805 as an average), $Al_2O_3$ - 1440, 1570, 1440,

1410, 1380 (for different adsorption bands), 1448 as an average, $SiO_2$ - 1440, $Y_2O_3$ - 1470, $TiO_2$ (anatase) - 1310, 1340, 1380 (1343 s an average), $TiO_2$ (rutile) - 1375, $ZrO_2$ - 1495, 1600 (1547 as an average), $P_2O_5$ - 1300. In general, the changes in the acidity of OH groups on oxide surface follow the regularities governed by the Periodic law: the higher is the position occupied by an element in Mendeleev's Table (for each group), the higher is proton donor ability of the surface. For each period, the acidity of groups increases with increasing of element number.

For quantitative estimations of acid-base properties, the above-described ideas of electronegativity are used. Besides electronegativities of atoms, Table 2.2 shows also the data of orbital electronegativities of the oxides and hydroxides. These data suggest that small values characterize basic properties while large values refer to acidic properties; intermediate values are typical for compounds with amphoteric properties.

Different researchers tried to establish a correlation between acid-base properties of compounds, in particular, of oxides, and their thermodynamic parameters, in order to range them in a definite row and to get a scale representing quantitative characterization of acid-base properties.

One of such scales has been proposed in [23]. This scale is based on Gibbs energy of the formation of sulphates from a basic oxide and sulphuric oxide $SO_3$, and of sodium salts from acid and sodium oxides. The data are shown in Table 2.4.

*Table 2.4.* The row of basic oxides according to the decrease of the energy of sulphate formation and the increase of the energy of sodium salt formation [23].

| Basic oxide | $\Delta G_2^0$ of sulphate, kJ/mol | Acidanion | $\Delta G_{form}^0$ of sodium salts, kJ/mol |
|---|---|---|---|
| $Cs_2O$ | -328,1 | $ZnO_2^{2-}$ | -11,3 |
| RbO | -323,5 | $OH^-$ | -71,9 |
| $K_2O$ | -311,8 | $SnO_3^{2-}$ | -79,4 |
| BaO | -227,0 | $FeO_2^-$ | -79,4 |
| SrO | -201,9 | $PbO_3^{2-}$ | -83.6 |
| $Li_2O$ | -195,2 | $AlO_2^-$ | -89,9 |
| CaO | -173,0 | $TiO_3^{2-}$ | -96,1 |
| $Tl_2O$ | -152,6 | $NO_2^-$ | -110,8 |
| $La_2O_3$ | -129,6 | $SbO_4^{3-}$ | -120,0 |
| PbO | -125,8 | $SiO_3^{2-}$ | -122,5 |
| $ZrO_2$ | -125,0 | $CO_3^{2-}$ | -138,4 |

| | | | |
|---|---|---|---|
| $Nd_2O_3$ | -120,8 | $AsO_4^{3-}$ | -149,2 |
| $Sm_2O_3$ | -117,0 | $MoO_4^{2-}$ | -153,8 |
| $Ag_2O$ | -117,0 | $WO_4^{2-}$ | -164,3 |
| MgO | -116,6 | $VO_3^-$ | -164,7 |
| CdO | -112,0 | $Mo_2O_7^{2-}$ | -170,1 |
| MnO | -111,0 | $CrO_4^{2-}$ | -168,9 |
| $In_2O_3$ | -103,2 | $PO_4^{3-}$ | -182,7 |
| $Hg_2O$ | -102,4 | $SO_3^{2-}$ | -188,1 |
| $Bi_2O_3$ | -100,3 | $IO_3^-$ | -193,1 |
| FeO | -99,5 | $Cr_2O_7^{2-}$ | -193,1 |
| NiO | -93,6 | $NO_3^-$ | -238,3 |
| ZnO | -91,5 | $PO_3^-$ | -246,2 |
| CoO | -89,0 | $SO_4^{2-}$ | -252,9 |
| HgO | -83,6 | $ClO_4^-$ | -356,5 |
| CuO | -82,3 | | |
| $Mn_2O_3$ | -76,9 | | |
| BeO | -71,7 | | |
| $Al_2O_3$ | -66,5 | | |
| $SnO_2$ | -36,8 | | |

The acidity of cation changes according to its oxidation degree and the number of the surrounding ions, i.e. its coordination number. For example, when the oxidation degree for manganese ion increases by unit, the acidity increases by 2-3 times, while for silicon ion $Si^{4+}$, it decreases when coordination number increases from 4 to 6.

The influence of the mentioned parameters on acid-base properties can be considered on the basis of force characteristics of ions, which, unlike electronegativities, are more sensitive to the electron structure of elements [24]. This is explained by the fact that the equation for the force characteristics $\gamma_i$, determined as a ratio of n-ionization potential $I_n$ to efficient radius $r_i$,

$$\gamma_i = I_n/r_i \qquad (2.24)$$

includes ion radius, which depends on the electron structure of element.

It was found out that the strength of cation binding to the $O^{2-}$ anion in oxygen-containing compounds is better expressed by the effective characteristic, i.e., $\gamma_i$ divided by coordination number, which defines the cation – ligand force in every direction of the coordination polyhedron. This characteristic is an analog of electronegativity, since it is also calculated per one bond. It is based on experimental data ($I_n$, $r_i$, coordination numbers).

Table 2.5 shows these characteristics as they increase. One can see that there is an agreement with the data shown in Tables 2.2 and 2.4. The first assumptions concerning the possibility of reaction to proceed in mixtures of components with different acid-base properties can be made using these data. It is evident that the highest efficiency should be expected in mixtures exhibiting a maximum difference in acid-base properties of components.

*Table 2.5.* Values of $\gamma_i$ divided by coordination number (an analogue of electronegativity) for some cations, ranged as the value increases [26,27,28].

| Cation | Coordina-tion number | Ionization potential $I_n$, eV | Ion radius, nm | $\gamma = I_n/r_i$ | γ/coordina-tion number |
|---|---|---|---|---|---|
| $Cs^+$ | 12 | 3.89 | 0.202 | 19.25 | 1.60 |
| | 8 | | 0.196 | 19.85 | 2.48 |
| | 6 | | 0.184 | 21.14 | 3.52 |
| $Rb^+$ | 12 | 4.18 | 0.187 | 22.35 | 1.86 |
| | 8 | | 0.174 | 24.02 | 3.00 |
| | 6 | | 0.163 | 25.64 | 4.28 |
| $K^+$ | 12 | 4.34 | 0.174 | 24.94 | 2.08 |
| | 8 | | 0.165 | 26.30 | 3.29 |
| | 6 | | 0.152 | 34.96 | 5.83 |
| $Na^+$ | 8 | 5.14 | 0.130 | 39.54 | 4.94 |
| | 6 | | 0.116 | 44.31 | 7.38 |
| | 4 | | 0.113 | 45.49 | 11.37 |
| $Li^+$ | 6 | 5.39 | 0.088 | 61.25 | 10.21 |
| | 4 | | 0.073 | 73.83 | 18.46 |
| $Ba^{2+}$ | 12 | 10.00 | 0.174 | 57.47 | 4.79 |
| | 8 | | 0.156 | 64.10 | 8.01 |
| | 6 | | 0.150 | 66.66 | 11.11 |
| $Sr^{2+}$ | 12 | 11.03 | 0.154 | 71.62 | 6.49 |
| | 8 | | 0.139 | 79.35 | 9.92 |
| | 6 | | 0.127 | 86.85 | 14.47 |

| | | | | | |
|---|---|---|---|---|---|
| $Ca^{2+}$ | 12 | 11.87 | 0.149 | 79.66 | 6.64 |
| | 8 | | 0.126 | 94.21 | 11.78 |
| | 6 | | 0.114 | 104.12 | 17.35 |
| $Pb^{2+}$ | 12 | 15.03 | 0.163 | 92.21 | 7.68 |
| | 8 | | 0.145 | 103.65 | 12.96 |
| | 6 | | 0.132 | 113.86 | 18.98 |
| | 4 | | 0.108 | 139.17 | 34.79 |
| $Sn^{2+}$ | 8 | 14.63 | 0.136 | 107.64 | 13.45 |
| $Mn^{2+}$ | 8 | 15.64 | 0.107 | 146.17 | 18.27 |
| | 6(H)* | | 0.097 | 161.24 | 26.87 |
| | 6(L)** | | 0.081 | 193.09 | 32.18 |
| $Mg^{2+}$ | 8 | 15.04 | 0.103 | 146.00 | 18.25 |
| | 6 | | 0.086 | 174.88 | 29.15 |
| | 4 | | 0.072 | 208.89 | 52.22 |
| $Fe^{2+}$ | 6(H) | 16.18 | 0.092 | 175.87 | 29.31 |
| | 6(L) | | 0.075 | 215.73 | 35.95 |
| | 4 | | 0.077 | 210.13 | 52.53 |
| $Co^{2+}$ | 6(H) | 17.06 | 0.088 | 193.86 | 32.30 |
| | 6(L) | | 0.079 | 215.95 | 35.99 |
| | 4 | | 0.071 | 240.28 | 60.07 |
| $Ni^{2+}$ | 6 | 18.15 | 0.083 | 218.67 | 36.44 |
| $Zn^{2+}$ | 8 | 17.96 | 0.104 | 172.69 | 21.5 |
| | 6 | | 0.089 | 201.79 | 33.63 |
| | 4 | | 0.074 | 242.70 | 60.67 |
| $Cu^{2+}$ | 6 | 20.29 | 0.087 | 233.22 | 38.87 |
| | 4 | | 0.176 | 266.97 | 66.74 |
| $La^{3+}$ | 12 | 19.17 | 0.146 | 131.30 | 10.94 |
| | 8 | | 0.132 | 145.22 | 18.15 |
| | 6 | | 0.119 | 161.09 | 26.85 |
| $Y^{3+}$ | 8 | 20.5 | 0.115 | 178.26 | 22.28 |
| | 6 | | 0.104 | 197.11 | 32.85 |
| $Bi^{3+}$ | 8 | 25.6 | 0.125 | 204.80 | 25.6 |
| | 6 | | 0.116 | 220.68 | 36.78 |
| $Cr^{3+}$ | 6 | 31.0 | 0.076 | 407.89 | 67.98 |

| | | | | | |
|---|---|---|---|---|---|
| $Mn^{3+}$ | 6(H) | 33.69 | 0.079 | 426.45 | 71.07 |
| | 6(L) | | 0.072 | 467.92 | 77.99 |
| $Fe^{3+}$ | 6(H) | 30.64 | 0.079 | 387.8 | 64.64 |
| | 6(L) | | 0.069 | 444.06 | 74.00 |
| | 4 | | 0.063 | 486.34 | 121.59 |
| $Ga^{3+}$ | 6 | 30.70 | 0.074 | 388.60 | 64.77 |
| | 4 | | 0.061 | 503.27 | 125.82 |
| $Al^{3+}$ | 6 | 28.44 | 0.067 | 424.48 | 70.75 |
| | 5 | | 0.062 | 458.71 | 91.74 |
| | 4 | | 0.053 | 536.60 | 134.15 |
| $Zr^{4+}$ | 8 | 33.97 | 0.092 | 369.23 | 46.15 |
| | 6 | | 0.086 | 395.0 | 65.84 |
| $W^{4+}$ | 6 | 35.0 | 0.079 | 443.0 | 73.83 |
| $Ti^{4+}$ | 6 | 43.24 | 0.075 | 576.53 | 96.09 |
| $V^{4+}$ | 6 | 48.0 | 0.073 | 657.53 | 109.58 |
| $Ge^{4+}$ | 6 | 45.7 | 0.068 | 761.66 | 126.94 |
| | 4 | | 0.054 | 846.30 | 211.57 |
| $Si^{4+}$ | 4 | 45.13 | 0.040 | 1120.82 | 280.20 |
| $V^{5+}$ | 6 | 56.0 | 0.068 | 823.53 | 137.25 |
| | 4 | | 0.050 | 1120.0 | 280.00 |
| $W^{6+}$ | 6 | 61.0 | 0.074 | 824.32 | 137.38 |
| | 4 | | 0.056 | 1089.28 | 272.32 |
| $P^{5+}$ | 4 | 65.1 | 0.031 | 2100.0 | 525.00 |
| $N^{5+}$ | 3 | 47.43 | 0.020 | 2371.5 | 790.50 |

* (H) – high-spin state, ** (L) – low-spin state.

### 2.2.2. The solubility of hydroxides

Water is one of the products formed via mechanochemical interaction between hydroxides. Its presence can substantially effect on the mechanism of transport processes and, thus, on the kinetics of mechanochemical processes, since some reaction products (initial or final) can be dissolved in it. Hence, the reaction could proceed with the participation of the dissolved forms of components.

The major part of hydroxides are poorly soluble or insoluble in water, except well-soluble hydroxides of alkaline metals and thallium, and much worse soluble hydroxides of alkaline earths.

The analysis of the solubility of compounds is usually made on the basis of the solubility product (L), which, in fact, is the equilibrium constant of the process

$$A_xB_{y(solid)} = xA^{y+}{}_{(aq)} + yB^{x-}{}_{(aq)}. \quad (2.25)$$

$$\text{Thus, } L = [A^{y+}]^x\ [B^{x-}]^y/[A_xB_y]. \quad (2.26)$$

The equilibrium constant is connected with free energy of dissolution $\Delta G$ by the equation

$$\Delta G_s = -\ RT\ln L. \quad (2.27)$$

The state of ions in an infinitely diluted solution (ion activity product) is selected to be the standard state, and the activity of $A_xB_{y(solid)}$ is taken to be unity.

It is admitted to use the concentrations expressed in moles per 1000 g of $H_2O$ (molality), instead of activities for poorly soluble compounds, and the index $pL = -\ \lg L$, instead of concentrations product.

For hydroxides possessing partially covalent bond, Clifford discovered a linear dependence between the difference in electronegativities $\Delta X$ of A and B (in the case when $\Delta X$ is not very large) and the solubility [29].

For the hydroxides of divalent metals (Fe, Co, Ni, Mg, Ca, Mn, Cu, etc.), an equation using the least squares was proposed (here, $\Delta X$ according to Pauling):

$$pL = 65.9 - 24.4\ \Delta X. \quad (2.28)$$

The largest deviations from linearity are observed for $Be(OH)_2$ and $Sn(OH)_2$.

A similar correlation for the hydroxides of trivalent metals (Ce, La, Y, Sc, Al, Cr, In, Ga, Fe, Co, Tl, Ir, Au) brings to the following equation

$$pL = 86.1 - 27.4\ \Delta X. \quad (2.29)$$

In this case, the deviation is the largest for $Au(OH)_3$; however, the accuracy of its measured solubility is doubtful.

We obtained a satisfactory correlation between the solubility index pL and the parameter $\gamma$/coordination number, proposed by Godovikov [24]. This parameter is an analogue of electronegativity. We accepted coordination number equal to 6. The data are listed in Table 2.6. Hydroxides that deviate from the correlation are those of $Pb^{2+}$ and $Sn^{2+}$; they are related to the intermediate acids, according to Pearson's classification; the others are related to hard acids. The reason of this deviation, probably, consists in their soft character.

*Table 2.6.* Correlation between the solubility product of hydroxides and $\gamma_i$ value divided by coordination number 6.

| Hydroxide | -lgL=pL [30] | γ/coord. number |
|---|---|---|
| CsOH | -13.75 | 3.52 |
| RbOH | -12.50 | 4.28 |
| KOH | -10.70 | 5.83 |
| NaOH | -7.20 | 7.38 |
| $Ba(OH)_2$ | -3.30 | 11.11 |
| LiOH | -1.87 | 10.21 |
| $Sr(OH)_2$ | -0.37 | 14.47 |
| $Ca(OH)_2$ | +5.00 | 17.35 |
| $Mg(OH)_2$ | +10.95 | 29.15 |
| $Mn(OH)_2$ | +12.80 | 32.18 |
| $Cd(OH)_2$ | +14.00 | 25.80 |
| $Fe(OH)_2$ | +15.15 | 35.95 |
| $Co(OH)_2$ | +15.40 | 35.49 |
| $Zn(OH)_2$ | +16.83 | 33.63 |
| $Ni(OH)_2$ | +17.82 | 36.44 |
| $Cu(OH)_2$ | +19.38 | 38.87 |
| $Pb(OH)_2$ | +19.70 | 18.98 |
| $La(OH)_3$ | +19.70 | 26.85 |
| $Y(OH)_3$ | +19.50 | 32.85 |
| $Sn(OH)_2$ | +25.70 | 13.45 |
| $Cr(OH)_3$ | +29.5 | 67.98 |
| $Al(OH)_3$ | +33.6 | 70.75 |
| $Ga(OH)_3$ | +37.2 | 64.80 |
| $Mn(OH)_3$ | +37.6 | 77.90 |
| $Fe(OH)_3$ | +38.6 | 74.00 |
| $Co(OH)_3$ | +40.5 | 74.42 |

| | | |
|---|---|---|
| $Sn(OH)_4$ | +56.0 | 81.79 |
| $HWO_4$ | +21.62 | 137.38 |

### 2.2.3. Thermal properties

Hydroxides and hydrated oxides lose water under heating and turn into anhydrous ordinary oxides (except strongly basic hydroxides of alkaline metals that melt without decomposition).

The data on decomposition temperatures of hydroxides and hydrated oxides, as well as the data on enthalpy and entropy of their formation, are listed in Table 2.7 according to [30-32].

*Table 2.7.* Decomposition temperature, enthalpy and entropy of formation for some hydroxides and hydrated oxides [30-32].

| Compound | Temperature range of water removal, °C | $\Delta H^o_{form}$, kJ/mol | $S^o_{298}$, J/mol·K |
|---|---|---|---|
| LiOH | Melting 473 | -484.9 | 42.86 |
| LiOH $H_2O$ | 40-109 | -788.3 | 71.4 |
| $Mg(OH)_2$ | 350-480 | -924.35 | 63.2 |
| $Ca(OH)_2$ | 480-620 | -983.68 | 83.3 |
| $Mn(OH)_2$ | 145-230 | -701.4 | 81.5 |
| MnO(OH) | 360-400 | - | - |
| $Fe(OH)_2$ | 100-300 | -572.7 | 88.0 |
| α-FeO(OH) | 280-380 | -559.8 | - |
| $Co(OH)_2$ | 150-300 | -541.0 | 83.0 |
| CoO(OH) | 120-190 | - | - |
| $Ni(OH)_2$ | 280-380 | - | - |
| $Al(OH)_3$, gibbsite | 325-330 | -1293.5 | 68.4 |
| $Al(OH)_3$, bayerite | 305-310 | -1288.3 | 68.4 |
| $Al(OH)_3$, norstandite | 300-330 | - | - |
| AlOOH, boemite | 450-580 | -986.8 | 48.4 |

| | | | |
|---|---|---|---|
| AlOOH, diaspore | 540-585 | -999.0 | 35.2 |
| $La(OH)_3$ | | -1408.7 | 34.6 |
| $H_4SiO_4$ | | -1481.1 | 192.0 |
| $SiO_2{\cdot}nH_2O$, $H_2SiO_3$ | | -1188.7 | 134.0 |
| $H_2Si_2O_5$ | | -2088.7 | 192.0 |
| $H_6Si_2O_7$ | | -2669.4 | 331.0 |
| $H_2TiO_3$ | 60-470 | | |
| $H_2V_2O_6$ | 300-400 | | |
| $H_2WO_4$ | 180-300 | | |
| $H_3PO_4$ | | | |
| $H_3PO_4{\cdot}0.5H_2O$ | | -1283.0 | |

One can see that decomposition temperatures of these compounds are within temperature range 40-600°C. Some compounds, such as monohydrate of lithium hydroxide (40°C), hydrated titanium dioxide (60°C), iron hydroxide (100°C), manganese (145°C) and cobalt (150°C) hydroxides, tungsten acid (180°C), etc., start to release water at relatively low temperatures. Other compounds decompose at a temperature above 200°C. No correlation between formation enthalpy and thermal stability of hydroxides and hydrated oxides is observed.

## 2.3. The preparation of hydroxides

Hydrated oxides with clearly exhibited basic and acidic properties are well soluble in water. They are prepared by the interaction of the corresponding oxides or metals with water. Weak bases, ampholytes and weak acids are practically insoluble or poorly soluble. Exchange reactions between the solutions of the corresponding salts and strong bases are used to synthesize them.

The most pure precipitates of the mentioned hydroxides are obtained from nitrate solutions of relatively low concentrations (till 1 N) with a stoichiometric amount of the precipitating agent (sodium hydroxide or ammonia) [32].

The hydroxide precipitates as prepared always contain variable amounts of non-structural water (as $H_2O$ molecules) depending on precipitation conditions. Water content decreases when passing from diluted solutions to concentrated ones, and from slow (dropwise) addition of precipitating reagent to rapid (flow). The amount of non-structural water decreases with precipitate ageing. The water content of precipitates decreases when passing from nitrates to chlorides and sulphates.

The dispersity of hydroxide precipitates can be governed by changing precipitation conditions. Particle size increases in the following row of salts used for precipitation: sulphates – chlorides – nitrates, and with passing from fast precipitation from concentrated solutions to slow precipitation from diluted ones.

As a rule, the fresh precipitates are X-ray amorphous. Crystallization is accelerated at increased temperature of the mother solution; crystallization also takes place during precipitate ageing.

A fundamental phenomenon for the consideration of hydroxide formation process is a hydrolysis, which is an exchange decomposition of the compounds by water

$$M^{n+} + H_2O \rightarrow MOH^{(n-1)+} + H^+ \qquad (2.30)$$

characterized by the hydrolysis constant

$$K = \{[(MOH)^{(n-1)+}]\cdot[H^+]\}/\,[M^{n+}] \qquad (2.31)$$

This constant involves concentrations; the activity of water is accepted to be 1.

The ability of compound to be hydrolyzed depends on the cation parameters $I_n$ and $\chi$ which govern the polarizability of Me – O bond and the formation of strong hybrid orbitals. For d-elements, d- , s- and p-orbitals in the outer electron shells are close to each other by energy. Therefore, d-elements exhibit an increased trend to hybridization with the participation of d-orbitals. This fact, as well as large I and X values, help to form stable complex compounds, while the ability to hydrolysis is closely correlated with the activity to complex formation.

Hydrolysis results in the formation of hydroxo-complexes which exhibit more or less strong tendency to form polynuclear hydroxo-complexes as a result of polymerization or polycondensation processes. Complex-forming ions are bound to each other with bridge OH groups (ol-bonds) or O-bridges (oxo-bonds). The formation of these bonds is an essential part of the ageing processes of the solutions prone to hydrolysis. The higher is coordination activity of a cation, the stronger is its tendency to hydrolysis and to multinuclear hydroxo-complexes formation.

Table 2.8 shows hydrolysis constants for some cations and '$\gamma$ /coordination number 6' values taken from Table 2.3.

*Table 2.8.* A correlation between hydrolysis constants of cations and γ/coordination number 6 values.

| $M^{n+}$ | lgK | γ/coordination number 6 |
|---|---|---|
| $Na^{+}$ | -14.48 | 7.38 |
| $K^{+}$ | -13.82 | 5.83 |
| $Ba^{2+}$ | -13.40 | 11.11 |
| $Sr^{2+}$ | -13.18 | 14.47 |
| $Ca^{2+}$ | -12.70 | 17.35 |
| $La^{3+}$ | -10.70 | 26.85 |
| $Mg^{2+}$ | -11.42 | 29.15 |
| $Mn^{2+}$ | -10.70 | 32.18 |
| $Fe^{2+}$ | -10.10 | 35.95 |
| $Zn^{2+}$ | -9.60 | 33.63 |
| $Co^{2+}$ | -9.60 | 35.99 |
| $Ni^{2+}$ | -9.40 | 36.44 |
| $Pb^{2+}$ | -7.78 | 18.98 |
| $Cu^{2+}$ | -7.53 | 38.87 |
| $Y^{3+}$ | -7.00 | 32.85 |
| $Al^{3+}$ | -5.14 | 70.75 |
| $Sn^{2+}$ | -4.30 | 13.45 |
| $Cr^{3+}$ | -4.01 | 67.98 |
| $Ga^{3+}$ | -3.40 | 64.77 |
| $V^{3+}$ | -2.92 | 68.70 |
| $Fe^{3+}$ | -2.19 | 74.00 |
| $Co^{3+}$ | -2.00 | 50.31 |
| $Bi^{3+}$ | -1.58 | 36.78 |
| $Zr^{3+}$ | -0.22 | 65.84 |

A clear correlation with the mentioned values is observed for $Na^{+}$, $Li^{+}$, $Ba^{2+}$, $Sr^{2+}$, $Ca^{2+}$, $Mg^{2+}$, $Mn^{2+}$, $Fe^{3+}$, $Ga^{3+}$, $Cr^{3+}$, $La^{3+}$, $Al^{3+}$, $V^{3+}$, $Zr^{4+}$ cations.

For $Sn^{2+}$, $Pb^{2+}$, $Bi^{3+}$, $Co^{3+}$ cations, a deviation from the mentioned correlation is observed. According to Pierson's classification, the former cations are related to hard acids, while the latter ones are considered as intermediate ones, i.e., the reason of the deviation, probably, consists in their softness.

Detailed investigations on the formation of precipitates from the viewpoint of the existence of intermediate products during the formation of product particles have been carried out. For magnesium hydroxide, these investigations are described in [33].

Chemical bond Mg – OH is characterized by the predominance of the ionic constituent (65-80 %). In alkaline precipitation from magnesium hydroxide solutions, its ageing occurs rather rapidly according to the mechanism of collective recrystallization without changing of phase composition. This process is accompanied by a decrease of specific surface area [33]. Due to weak coordination ability of $Mg^{2+}$, no polynuclear hydroxo- complexes are formed in solutions as was demonstrated by means of NMR [34]. Therefore, the time of magnesium hydroxide deposition from solution is much shorter than that for other hydroxides.

$Al^{3+}$ ion is a typical complexing agent able to form strong complexes with tetrahedral $sp^3$ coordination bonds. The presence of $Al(H_2O)_6{}^{3+}$, $Al_2(OH)_2(H_2O)_4{}^{4+}$, $Al_{13}O_4(OH)_{24}{}^{7+}$ complexes in solutions was separately identified by means of NMR $^{27}Al$ [35]. The structure of these complexes was investigated by means of X-ray diffraction [36].

Hydroxo-complexes are formed via polycondensation mechanism including the stages of ligand substitution, olation (formation of Al-OH bonds) and oxolation (formation of Al-O bonds) [37]. When definite size of particles and supersaturation degree are achieved, the condensed particles grow together according to the fluctuation mechanism till the size of primary particles is achieved. The mechanism of the formation of primary particles is poorly investigated; however, it is known that they have spherical or oval form, their size being up to 1.0 - 4.0 nm. The aggregation of particles proceeds via coagulation mechanism which is stepwise. Weak coagulation bonds turn into condensation bonds with time that causes irreversible ageing of the system.

A large fraction of covalent character in Al-O bonds (63 %) provides a significant duration of all the processes of the formation of aluminium hydroxide via alkaline precipitation from salts solutions. According to NMR relaxation data, the formation of aluminium hydroxide sol proceeds from several tens minutes to hours, while aluminium hydroxide ageing under the layer of mother solution can proceed hours and days [38].

A substantial fraction of the ionic character of Al – O bond and the absence of the own electrons in the valence shell of $Al^{3+}$ make the complexes kinetically labile that allows to obtain crystal aluminium hydroxides (gibbsite, bayerite, boemite, norstandite and diaspore) using the reactions in water solutions [33,34].

The formation of oxyhydrate precipitates during the hydrolysis of silicon salts proceeds via the scheme:

I. Hydrolysis and formation of polymeric hydroxo-complexes

↓

II. Formation of primary particles

↓

III. Coagulation with the formation of gels

↓

IV. Ageing of precipitates

However, each of these stages has its specific features, which is explained by clearly exhibited properties of the $Si^{4+}$ ion possessing powerful force characteristics and high electronegativity. First of all, the polymerization of silicic acid proceeds via the anion type but not cation one. This process is accompanied by the formation of chains and nets of $SiO_4$ tetrahedrons which possess four strong Si-O δ-bonds formed similarly to $sp^3$ hybridization of valence orbitals of silicon. The formation of polymers occurs according to the condensation mechanism $(HO)_3SiOH + OHSi(OH)_3 \rightarrow (HO)_3SiO - O - Si(OH)_3 + H_2O$.

As a result of growing together, globules of primary particles of the size 2 – 100 nm are formed at the II stage. Due to strong orienting action of $Si^{4+}$ ion, the structure of the primary particle is an openwork carcass composed of $SiO_4$ tetrahedrons with a large number of the end OH groups and water molecules in the voids between the tetrahedrons [39]. Hydrogen ions of OH groups can participate in ion exchange reactions with cations.

Coagulation of hydrated silica sols at the III stage leads to the formation of globular gels in which primary particles are randomly packed. Primary particles which comprise coagulations products are bound at first by weak coagulation forces; later these turn into condensation bonds of the Si – O – Si type. In general, the processes that take place at the II and III stages of structure formation result in the formation of a framework carcass of $SiO_4$ tetrahedrons and their globular aggregates. Thus formed carcass possesses increased rigidity and acts against Laplas forces during the drying of silicic acid gels. Because of this, xerogels possess a developed specific surface (up to 800 $m^2/g$) and a system of ultra- and micro-pores.

At the IV stage, the ageing of gels is accompanied by the transformation of coagulation bonds into condensation bonds, and by re-condensation processes.

The processes which take place according to the mentioned stages are partially overlapped that brings to complications into the mechanism of silica gel formation and into the preparation of products with prescribed properties. In particular, it is impossible to obtain crystal silicic acid from gels under usual conditions.

The most up-to-date opinion concerning the behaviour of $Ti^{4+}$ ions in solutions is reported in [40, 41]. The mechanism of the formation of amorphous precipitates is considered within the frames of a scheme outlined above for silicic acid.

$Ti^{4+}$ ions can form multinuclear linear and voluminous complexes in solutions. The nuclei of these complexes are bound to each other via ol and oxol bonds. There is a connection between the structure of primary hydroxo-complexes and hydroxide gels. In particular, depending on the method of preparation of hydrated titanium dioxide, $TiO_2$ of either anatase or rutile structure can be obtained.

The energy states of 3d-, 4s- and 4p-orbitals being closed to each other, together with high ion potential, provide an increased tendency of $Ti^{4+}$ ions to form hybrid bonds $d^2sp^3$ with a strong polarization of $O^{2-}$ ions. A rigid orientation of coordination bonds of $Ti^{4+}$ ions with ligands brings to complications into rearrangement of the structures in solution and precipitates. The tendency to form intermediate structures is clearly exhibited by $Ti^{4+}$ ions that causes long induction periods, slow ageing of precipitates and solutions, amorphous character of hydroxide precipitates, etc.

The structure of titanium complexes affects the formation of hydrated titanium dioxide structure, since rutile and anatase lattices are composed of $TiO_6$ octahedrons connected in definite manner. The formation of anatase structure occurs when two octahedral complexes form a common vertex. When two octahedrons are united via their edges, rutile structure is formed. Based on this assumption, it is considered that if titanium (IV) complexes with one reactive centre are formed during hydrolysis, anatase structure is formed; if there are two reactive centres, then rutile structure is formed.

The results of fundamental investigations of the formation of hydrated zirconium dioxide are reported in [41]. The oxidation degree of zirconium ions is +4 in nearly all its compounds. Possessing rather large radius (0.092 nm) and rather low ion potential, it forms sterically maximal number of bonds. Accordingly, in the major part of its compounds, $Zr^{4+}$ ions exhibits coordination number 8.

This is also the evidence of a substantial fraction of ionicity of zirconium bonds with ligands. However, zirconium forms strong directed bonds. With coordination number 8, it uses hybrid p-orbitals of $d^4sp^3$ type, that leads to dodecahedral configuration of bonds. Thus, zirconium compounds are prone to hydrolysis and hydrolytic condensation. In particular, in the solutions of zirconium chloride $ZrOCl_2 \cdot 8H_2O$, a tetramer $[Zr_4(OH)_{16}(H_2O)_8]^{\circ}$ is the prevailing form, which is able to participate in further polymerisation processes. According to the data [42], the structure of the tetramer corresponds to the following scheme:

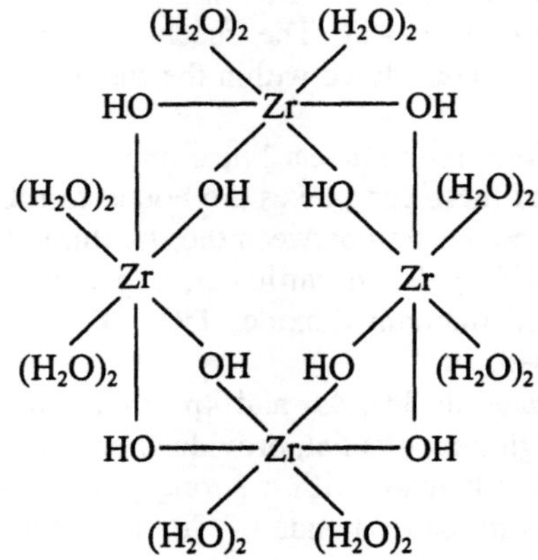


Along with olation, the solutions of $Zr^{4+}$ salts can also undergo oxolation processes. The presence of the anions of weak acids in solution helps this process. The addition of $F^-$ ions that get tightly bound to $Zr^{4+}$ and even are able, under definite conditions, to replace $OH^-$ groups from the inner coordination sphere, causes deolation and even deoxolation, thus, destroying polymer forms [43]. Vice versa, the addition of anions able to form strong bridge bonds with $Zr^{4+}$ ions strengthens ol bonds (the effect of mutual strengthening of bonds in conjugated bridges).

$Zr(OH)_4$ gels formed during hydrolysis and polymerization are composed of particles with a size of 3 nm, i.e. the formation of zirconium hydroxides occurs via the scheme similar to that describing the formation of hydrated silica gel.

The treatment of $Zr^{4+}$ solutions with alkaliis and ammonia leads to the formation of amorphous hydroxide which is stable against crystallization. The resistance to rearrangements is due to the nature of chemical bonds, their rigidity. This leads to the appearance of induction period in precipitation, to deceleration of precipitate ageing processes, etc. This is accompanied by high stability of zirconium complexes. Coordination number for $O^{2-}$ changes from 2 to 3-4 during dehydration of the amorphous $Zr^{4+}$ hydroxide. It is stated that this change allows to obtain intermediate oxide forms able to active agglomeration.

Thus, the mechanisms of processes which occur during the hydrolysis of many compounds are well explained from the viewpoint of coordination chemistry.

## *References*


1. Pauling, Linus, *The Nature of the Chemical Bond*. Pasadena, CA: Cornell University Press, 1939.
2. Mulliken R.S. Note on electric moment and infra-red spectra. A correction. J. Chem. Phys. 1934; 2: 712-13.
3. Hinze J., Jaffe H.H. Electronegativity. IV. Orbital electronegativities of the neutral atom of the period III A and IV A and of positive ions of periods I and II. J. Phys. Chem. 1963; 67: 1501-6.
4. Iczkowski R.P., Margrave J.L. Electronegativity. J. Amer. Chem. Soc. 1961; 83: 3547-51.
5. Emsley, John, *The Elements*. 2nd edition. Oxford: Claredon Press, 1991.
6. Viting, Leonid, *The High-Temperature Solutions – Meltings*. Moskva: MGU, 1991.
7. Sanderson R.T. Electronegativities in Inorganic Chemistry. J. Chem. Educ. 1954; 31: 2-17.
8. Klopman G.J. Electronegativity. J. Chem. Phys. 1965; 43, Pt 2: S214-29.
9. Parr R., Yang W. *Density Functional Theory of Atoms and Molecules*. Oxford: Claredon Press, 1989.
10. Parr R.G., Donnelly R.A., Levy M., Palke W.E. Electronegativity: the density functional viewpoint. J. Chem. Phys. 1978; 68(8): 3801-7.
11. Parr R.G., Pearson R.G. Absolute hardness: companion parameter to absolute electronegativity. J. Amer. Chem. Soc. 1983; 27(4):734-40.
12. Pearson R.G. Absolute electronegativity and hardness: application to inorganic chemistry. Inorg.Chem. 1988; 27(4): 734-40.
13. Pearson R.G. Recent advances in the concept of hard and soft acids and bases. J. Chem. Educ. 1987; 64(7): 561-7.
14. Wells, Alexander, *Structural Inorganic Chemistry*. 4th edition. Oxford: Claredon Press, 1986.
15. Mirkin L.I. *Handbook: X-ray Structural Analysis of Crystals*. Moskva: Izd. Phys. Math. Lit, 1961.
16. Brönsted I, Rec. Trav. Chim., 1923; 42: 718-25.
17. Lowry T., Strength of acids and bases. Chem. and Ind. 1923; 42: 43-51.
18. Lewis, G., *Valence and the Structure of Atoms and Molecules*, New York, 1923.
19. Day, Clyde and Selbin, Joel, *Theoretical Inorganic Chemistry*. New York: Reinhold Publ. Corp., 1962.
20. Hammet L.P., Deyrup A.J. The acidity functions of mixtures of sulfuric and perchloric acids with water. J. Am. Chem. Soc. 1932; 54: 2721-29.
21. Tanabe K. Solid asids and bases. In *Catalysis by Acids and Bases*. Ed. B. Imelink. Amsterdam: Elsevier, 1985.
22. Paukschtis, Evgenii, *Infrared Spectroscopy in Heterogeneous Acid-Base Catalysis*. Novosibirsk: Nauka, 1992.
23. Kireev V.A. Acid-base properties of oxides. Zhurn, Fiz. Khim. 1964; 38: 1881-92.
24. Godovikov A.A. Using electronegativities in the systematics of minerals and inorganic substances. Zhurn. Neorg. Khim. 1993; 38: 1468-82.
25. Shannon R.D. Revised effective ionic radii and systematic studies of interatomic distances in halides and chalcogenides. Acta Cryst. 1976; 32A: 751-67.
26. Shannon R.D., Prewitt C.T. Effective ionic radii in oxides and fluorides. Acta Cryst. 1969; 25: 925-46.
27. Godovikov, Alexandr, *Orbital Radii and Properties of Elements*. Novosibirsk: Nauka, 1977.
28. Clifford A.F. The Prediction of Solubility Product Constants. J. Amer. Chem. Soc., 1957; 79: 5404-7.
29. Kumok V.N., Kuleschova O.M., Karabin L.A. *Solubility Product Constants*. Novosibirsk: Nauka, 1983.

30. *Differential Thermal Analysis. V.1. Fundamental Aspects.* Ed. Mackenzie R. C. London and New York: Academic Press, 1970.
31. *Thermal Constants of Inorganic Substances.* Ed. Gluschko V.P. In 10 volumes. Moskva: VINITI, 1970-1985.
32. Chalii V. P. *Hydroxides of Metals.* Kiev: Naukova Dumka, 1972.
33. Dzisko, Vera and Karnaukhov, Anatolii and Tarasova, Djemma. *Physico-Chemical Basis of Synthesis of Oxide Catalysts.* Novosibirsk: Nauka, 1978.
34. Akitt J.W. $^{27}$Al Nuclear magnetic resoance studies of the hydrolisis and polimerisation of the hexa-aquo-aluminium (III) cation. J. Chem. Soc. Dalton Trans. 1972; 5: 604-9.
35. Iohanson G. The crystal structures of $Al_2(OH)_2(H_2O)_8(SO_4)_2 \cdot 2H_2O$ and $Al_2(OH)_2(H_2O)_8(SeO_4)_2 \cdot 2H_2O$. Acta Chem. Scand. 1962; 16: 103-20.
36. Berestneva Z.Ya., Kargin V.A. About formation mechanism of colloidal particles. Uspekhi Chimii, 1955; 24: 249-59.
37. Dzisco V.A., Ivanova A.S., Vischnyakova G. P. Formation of aluminium hydroxide at aqing. Kinetika i Kataliz, 1976; 17: 483-90.
38. Vysotsky Z.Z., Galinskaya V.I., Kolytchev V.I. et al. About the role of polymerization and depolymerization of silicic acid in the processes of formation and rearrangement of the gel carcass. In *Adsorption and Adsorbents.* Kiev: Naukova Dumka, 1972.
39. Pletnev R.N., Ivanin A.A., Kleshchev D.G. *Hydrated Oxides of the Elements of IV and V groups.* Moskva: Nauka, 1986.
40. Santacesaria E., Tonello M., Storti G. et al. Kinetics of titanium dioxide precipitation by thermal hydrolisis. J. Colloid Interface Sci. 1986; 111: 44-56.
41. Reinten Kh.T. Equipment, preparation and properties of hydrated zirconium dioxide. In *Structure and Properties of Adsorbents and Catalysts.* Ed. Linsen B.G. Moskva: Nauka, 1973, p. 332-83.
42. Muha G.M., Vauoghan P.A. Structure of the complex ion in aqueous solutions of zirconyl and hafnyl oxyhalides. J. Chem. Phys. 1960; 33: 194-99.
43. Godneva M.M., Motov D.L. *Chemistry of Titanium Subgroup. Sulphates and Their Solutions.* Leningrad: Nauka, 1980.

# Chapter 3

# SOME THEORETICAL ASPECTS OF MECHANOCHEMICAL REACTIONS

## 3.1. Thermodynamics of acid-base reactions

Chemical interaction between solids involves the consumption of some compounds and the formation of others. These processes occur in agreement with physicochemical laws and can be characterized using the fundamental thermodynamic equations. The main of them is the equation

$$\Delta G = \Delta H - T\Delta S \qquad (3.1)$$

which describes the changes of the isobar-isothermal potential (Gibbs free energy) with the process enthalpy $\Delta H$ and entropy $\Delta S$, the functions of the state.

A principal possibility for one or another chemical reaction to occur is determined by the sign of $\Delta G$. According to the laws of thermodynamics, the process occurs spontaneously in the direction of decreasing the free Gibbs energy. A reaction between the substances proceeds with the formation of products, if

$$\Delta G_{reac} = \Sigma\Delta G_{final\ products} - \Sigma\Delta G_{initial\ products},\ \Delta G < 0 \qquad (3.2)$$

and takes the back direction if $\Delta G > 0$.
For example, in the reaction $A_{solid} + B_{solid} = AB_{solid}$, Gibbs energy is estimated as

$$\Delta G_{reac} = \Delta G^0_{reac} + RT \ln a_{AB}/(a_A \cdot a_B) \qquad (3.3)$$

where $a_A$, $a_B$, $a_{AB}$ are the activities of the initial and final products. However, since the activity of solids is equal (or close) to unity at normal pressure and room temperature, then

$$\Delta G_{reac} = \Delta G^0_{reac} \qquad (3.4)$$

Thus, in order to determine the possibility for a reaction to proceed, it is sufficient to know $\Delta G^0_{reac}$ in standard state ($p = 0.1$ MPa, $T = 298$ K) which is calculated as the difference between Gibbs energies of the formation of final and initial products in standard state. These values are listed in reference books.

If a reaction in a mixture of solids is accompanied by the formation of gas or fluid phases (melts, solutions), solid solutions, or by the generation of defects, then, for a more strict thermodynamic forecast, it is necessary to take into account the changes of entropy and specific heat capacity during phase transitions of the components (melting, vaporization, dissolution), changes of volume and other parameters. If these factors are not taken into account, one can come across the contradictions between experimental data and thermodynamic calculations.

It is known that the more is the difference in acid-base properties of the reagents, the faster and more complete is the interaction. Thermodynamic data confirm this statement. The Gibbs energies for the reactions of a similar type correlate with the differences in acid-base properties of these reagents. The larger is the difference, the more negative is the value of Gibbs energy. Table 3.1 represents the data on the formation of complex oxides from ordinary ones [1]. The oxides of elements from the upper left corner of the table which have the largest differences in acid-base properties are exhibited by more negative values of $\Delta G_{reac}$ too.

*Table 3.1.* Gibbs energy (kJ/g-equiv.) for the formation of compounds from ordinary oxides [1].

| | $SO_3$ | $N_2O_5$ | $CO_2$ | $SiO_2$ | $Fe_2O_3$ | $Al_2O_3$ |
|---|---|---|---|---|---|---|
| $K_2O$ | -311.8 | -290.5 | -173.5 | -131.7 | - | - |
| $Na_2O$ | -255.8 | -237.8 | -138.3 | -122.5 | -79.4 | -89.4 |
| BaO | -227.0 | -194.4 | -108.3 | -78.8 | -48.5 | -53.9 |
| SrO | -201.9 | -170.1 | -91.5 | -65.6 | - | -44.3 |
| $Li_2O$ | -195.2 | -163.4 | -91.1 | -71.4 | -59.3 | -56.8 |
| CaO | -173.0 | -129.6 | -64.8 | -44.3 | -33.4 | -11.7 |
| PbO | -125.8 | -91.9 | -23.4 | -8.4 | - | - |
| $Ag_2O$ | -117.0 | -87.8 | -15.9 | - | - | - |
| MgO | -116.6 | -70.6 | -32.2 | -18.0 | -26.3 | - |
| CdO | -112.0 | -75.2 | -25.5 | -10.4 | - | - |
| MnO | -111.2 | -133.7 | -30.1 | -41.8 | -8.4 | - |
| FeO | -99.5 | - | -17.3 | -9.2 | -14.6 | -20.1 |
| NiO | -95.6 | -69.0 | -4.3 | - | - | - |
| ZnO | -91.5 | -43.9 | -11.3 | - | -14.6 | - |
| CoO | -89.0 | -71.0 | -18.4 | - | - | - |

| | | | | | | |
|---|---|---|---|---|---|---|
| CuO | -82.3 | -52.2 | -2.5 | - | - | - |
| BeO | -71.0 | - | - | - | - | - |

It was mentioned above that the concept of average orbital electronegativity is used to characterize acid-base properties. According to this concept, acidity increases with increasing the numerical value of electronegativity while the basicity decreases, correspondingly.

We calculated the differences between average orbital electronegativities for some combinations of oxides listed in Table 3.1. The results are represented in Table 3.2.

*Table 3.2.* Difference between average orbital electronegativites for combinations of oxides.

| | $SO_3$ | $N_2O_5$ | $O_2$ | $SiO_2$ | $Fe_2O_3$ | $Al_2O_3$ |
|---|---|---|---|---|---|---|
| $K_2O$ | -3.14 | -3.37 | -3.06 | -2.55 | -2.14 | -1.76 |
| $Na_2O$ | -2.89 | -3.12 | -2.81 | -2.30 | -1.89 | -1.51 |
| BaO | -2.25 | -2.48 | -2.17 | -1.66 | -1.25 | -0.87 |
| SrO | -2.35 | -2.58 | -2.28 | -1.76 | -1.35 | -0.97 |
| $Li_2O$ | -2.69 | -2.92 | -2.61 | -2.10 | -1.69 | -1.31 |
| CaO | -2.39 | -2.62 | -2.31 | -1.80 | -1.39 | -1.01 |
| PbO | -1.32 | -1.55 | -1.24 | -0.73 | -0.32 | +0.06 |
| $Ag_2O$ | -1.73 | -1.96 | -1.65 | -1.14 | -0.73 | -0.35 |
| MgO | -1.53 | -1.76 | -1.55 | -0.96 | -0.53 | -0.15 |
| CdO | -1.25 | -1.48 | -1.17 | -0.66 | -0.25 | +0.13 |
| MnO | -1.77 | -2.00 | -1.69 | -1.18 | -0.77 | -0.39 |
| FeO | -1.31 | -1.54 | -1.23 | -0.72 | -0.31 | +0.01 |
| NiO | -1.20 | -1.43 | -1.12 | -0.61 | -0.20 | +0.18 |
| ZnO | -1.06 | -1.29 | -0.98 | -0.47 | -0.06 | +0.32 |

The comparison between Table 3.1 and Table 3.2 shows that the Gibbs energy and the difference of electronegativities correlate with each other. Fig. 3.1 shows this dependence for the reactions of sulphates and silicates formation.

For the synthesis of multicomponent compounds, when more than two components are present in the initial mixture, the correlation between $\Delta G_{reac}^{0}$ and $\Delta X = X_B - X_A$ is not valid. It can be assumed that for multicomponent mixtures, the probability of the formation of complicated compounds from simple ones will increase with the increasing of difference between the sums of average orbital electronegativities for the initial and final products, similarly to thermodynamic equation (3.2), i.e.,

$$\Delta X = \Sigma X_{fin.prod.} - \Sigma X_{init.prod.} \quad (3.5)$$

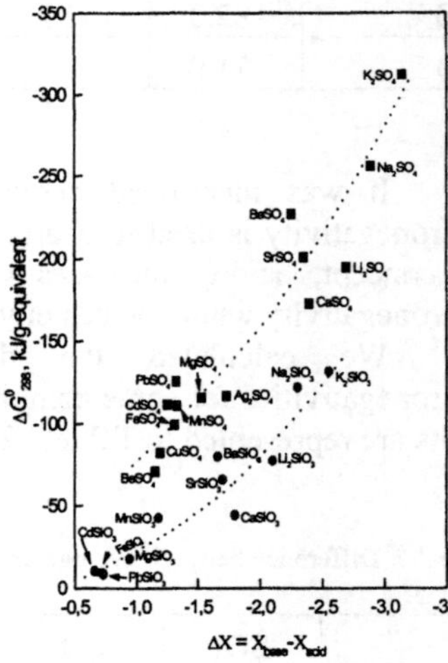


*Figure 3.1.* Correlation between Gibbs energies and differences of average orbital electronegativities of the initials oxides for the synthesis reactions of sulphates and silicates.

In order to test this statement, the calculation of Gibbs energies of the formation of complex oxides from ordinary ones on the basis of data taken from handbook [2] and the calculation $\Delta X$ according to the equation (3.5), based on absolute electro-negativities, listed above (see Table 2.1), were carried out.

According to Fig. 3.2, a good correlation exists between these

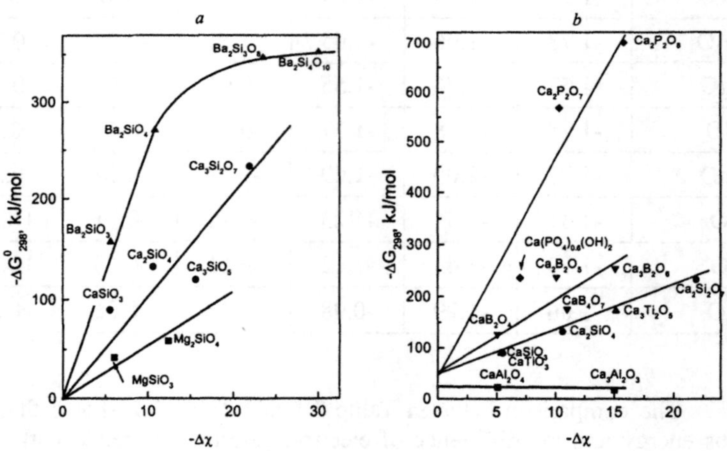


*Figure 3.2.* Correlation between Gibbs energy and differences of sum of average electronegativities (equation 3.5) for the synthesis of complex oxides.
*a* – alkaline earth silicates (common anion), *b* – calcium compounds (common cation)

values [3]. Using this correlation, one can estimate $\Delta G_{reac}$ for the reactions when thermodynamic data are unknown. The lack of thermodynamic data is typical, in particular, for multi-component compounds.

## 3.2. The kinetics of mechanochemical reactions

Thermodynamic considerations may be insufficient to make a conclusion on essential occurrence of a reaction if the conditions determining its kinetics are not taken into account.

The large value of activation energy in Arrhenius equation

$$k = A\, e^{-E/RT} \qquad (3.6)$$

where k is a reaction rate constant, A is a pre-exponential factor, E is an activation energy, R is a constant, and T is a temperature, a poor mixing of the components, a small particle contact area can make thermodynamically probable reaction impossible to occur.

The main statements of the kinetics of solid-phase reactions which bind the interaction rate with the diffusion of one of components, at first over the surface of another component, and then in the bulk through product layer, are not suitable for mechanochemical reactions, since mechanical activation of a mixture involves continuous renewal of contact surfaces. On this grounds, it was assumed that the interaction rate is limited by the rate of chemical reaction at contact.

Due to the pulsed action of balls on the reacting mixture, the process non-uniformity as in the time and in the space of a single mechanical action (the existence of gradient from the center of collision to its periphery), and changes of the conditions for chemical interaction in the course of activation, kinetic description of mechanochemical reactions is a complicated task. Therefore, one should not expect the creation of some universal reaction model; only some particular models are possible.

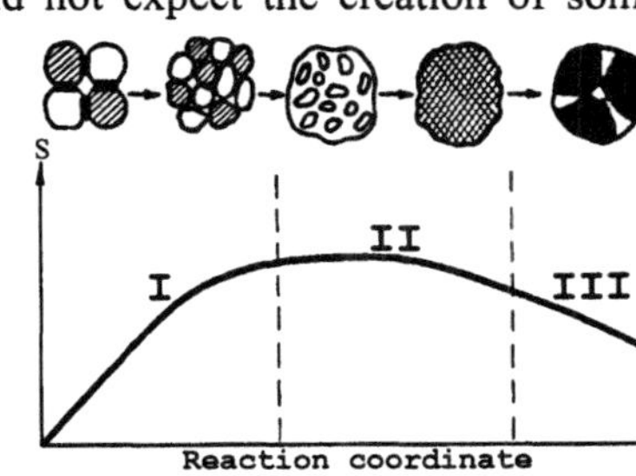


*Figure 3.3.* The change of the conditions for chemical interaction in the course of mechanical activation.
*S* – specific surface area.

Mechanical activation of mixtures involves the changes in the conditions for chemical interaction, as it follows from Fig. 3.3 [4,5]. The first stage of interaction is connected with the progressive growth of surface area during activation. Real solids are polycrystals or they contain the nuclei of microcracks. Therefore, their destruction occurs via separation into crystallites. The reaction proceeds at the contacts of particles.

At the second and third stages, the processes involving plastic deformation of particles are developed. The smaller is particle size, the more efficient are these processes. Dispersion process is overlapped by the formation of secondary particles, while the rate of the latter process is comparable with dispersing rate; thus, the surface area remains constant. Chemical reactions take place inside secondary aggregates at the contacts between particles. At the third stage, the crystallization of the products from the solid phase may occur, as well as its repeated amorphization, till some stationary state between these two is achieved.

The relations between the duration of stages depend on the amount of mechanical energy loading. If it is low, the process can stop at the second stage; at high energies, the second and third stages occur.

Since dispersing process and activation develop as statistically probable process, this allows to consider chemical interaction from the viewpoint of collision theory, but in this case, instead of binary collisions of molecules, as in gases or liquids, group collisions of surface atoms of two different particles in the zone of mechanical action are considered. The limiting stage of this process is the probability of formation of contacting regions between solid particles of the components and a collision with a milling body.

According to the mentioned statements, the rate of reaction in an operating grinding device can be estimate as

$$v = K_m x S_n, \tag{3.7}$$

where $K_m$ is a constant characterizing the probability of reaction to occur at a given mechanical action per unit contact area (energy constant determining thermodynamic reaction parameters T and p), x is a probability of contacting particles to come across a collision with milling body, $S_n$ is an area of contacting regions of A and B components during mechanical action [5].

Dispersion brings to an increase of $S_n$, since the surfaces of interacting components increase. It is known that the dependence of specific surface of a substance versus dispersion time obeys the equation:

$$S_t = S_m (1 - e^{-kt}), \tag{3.8}$$

where $S_m$ is a maximum specific surface area achieved under the given conditions, and k is a dispersion constant [6]. It is assumed that the initial surface value $S_o$ is small and can be neglected.

Since, as it follows from simple geometric considerations, $S_n$ is proportional to the total surface of particles, a similar equation can be also written for the contact surface:

$$S_n = S_{nm} (1 - e^{-kt}). \qquad (3.9)$$

On the other hand, as the reaction proceeds, the amount of initial reagents decreases; the probability for reagent particles to meet each other decreases. Similar to the law of mass action, we assume that the interaction rate can be written as

$$d\alpha / dt = K_m x\, S_n (1-\alpha)(1-\varepsilon\alpha), \qquad (3.10)$$

where $\alpha$ and $\varepsilon\alpha$ are the transformation degrees for substances A and B, respectively, $(1-\alpha)$ and $(1-\varepsilon\alpha)$ are the amounts of unreacted substances A and B, $\varepsilon = n/m$ is a coefficient of a mixture composition, where n is a stoichiometric coefficient representing the ratio of A to B moles in the product, and m is a real molar ratio of A to B in the initial mixture.

Integration of the equation (3.10) taking into account the equation (3.9) brings to the equation

$$1/(\varepsilon-1) \ln(1-\alpha)/(1-\varepsilon\alpha) = K [ t - 1/K_2 (1-e^{-kt})], \qquad (3.11)$$

where

$$K = K_m x\, S_{nm}. \qquad (3.12)$$

At stoichiometric components ratio n = m, $\varepsilon = 1$, the integration gives the following equation:

$$\alpha / (1 -\alpha) = K [ t - 1/ K_r (1- e^{-kt})]. \qquad (3.13)$$

In the case when specific surface area is linearly dependent on grinding time (i.e., $S_t = kt$), the equation (3.10) transforms into:

$$\alpha / (1-\alpha) = K t^2. \qquad (3.14)$$

In the case when surface area of the mixture is not changed during the treatment, i.e., $S_t$ = const (it occurs when the initial substances are taken in highly disperse state or when a stationary state is achieved at which the substances are not ground but only are subjected to plastic flow), the integration of (3.10) gives an equation similar to that describing by bimolecular reaction:

$$\alpha / (1-\alpha) = Kt. \qquad (3.15)$$

But the physical mean of the constant is quite different from that in molecular kinetic theory of chemical reactions.

As it was demonstrated in [7], a large number of papers on the kinetics of mechanochemical reactions report S-shaped experimental curves. The equations 3.11 and 3.12 provide a good description of this S-like shape.

Experimental curves obtained for mechanochemical reaction (without heating) $BaCO_3 + WO_3 = BaWO_4 + CO_2 \uparrow$ demonstrated a satisfactory agreement with the experiment [5]. Fig. 3.4*a* shows experimental data for this reaction, while Fig. 3.4*b* shows these data treated in the coordinates of the equation (3.14). The equation provides a good agreement with experiment till the transformation degree is less than 0.7. For other reactions, good agreement with experiments is also observed [5].

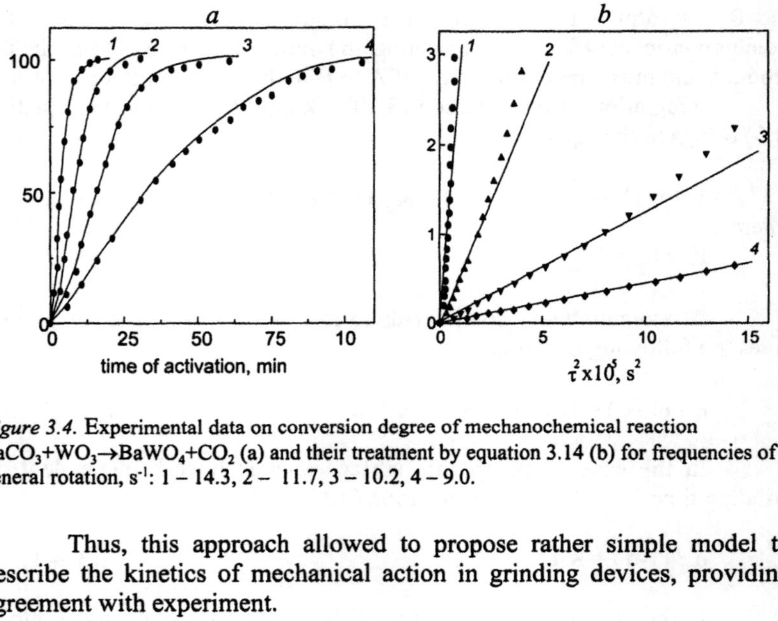


*Figure 3.4.* Experimental data on conversion degree of mechanochemical reaction $BaCO_3+WO_3 \rightarrow BaWO_4+CO_2$ (a) and their treatment by equation 3.14 (b) for frequencies of general rotation, $s^{-1}$: 1 – 14.3, 2 – 11.7, 3 – 10.2, 4 – 9.0.

Thus, this approach allowed to propose rather simple model to describe the kinetics of mechanical action in grinding devices, providing agreement with experiment.

## 3.3. Energy yield of mechanochemical reactions

One of the main problems in investigation of mechanochemical transformations consists in the relations between product yield and mechanical energy consumed by the process. Butyagin and Pavlychev [8] proposed to characterize mechanochemical yield by the ratio of the moles of product to the amount of the energy consumed (mol/MJ), similarly as in radiation chemistry. In reality, the researches most often record the dependence of the transformation degree $\alpha$ versus the time of mechanical treatment of powder mixture in mills. If the power of apparatus is known, the treatment time can be easy transformed into a dose: D = Wt, where W is a power of the apparatus per unit mass of material under treatment, and t is a treatment time.

The calculation of mechanochemical yields allows to compare the efficiencies of different devices. Besides, the comparison between data obtained at different amounts of consumed energy allows to conclude that mechanochemical yield per unit energy is a constant [7].

We obtained the data on mechanochemical yields for the reaction $BaCO_3 + WO_3 = BaWO_4 + CO_2\uparrow$ depending on power input. The power of a planetary mill was controlled by a frequency of rotation and determined experimentally from the consumption of electrical energy used only for balls operation (blank consumption being subtracted).

The results are shown in Fig. 3.5. According to Fig. 3.5*b*, mechanochemical yield per unit of consumed energy appeared to be a constant only starting from some threshold power per 1 g of the mixture under treatment. This threshold is about 5 W/g. The existence of threshold effect in solids is connected with fragile-viscous transition; the latter occurs only when a definite level of energy input to a solid is exceeded.

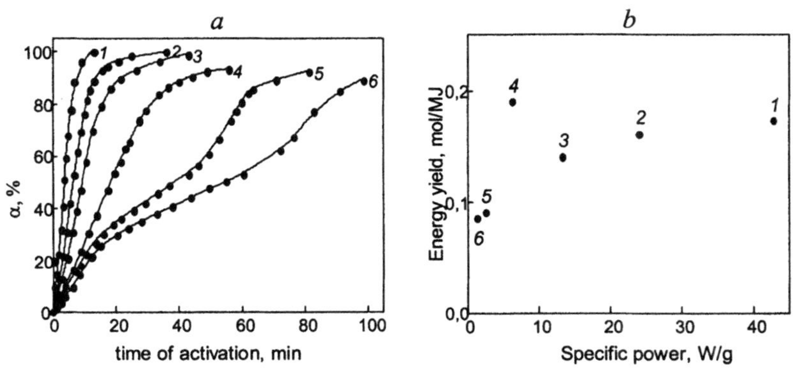


*Figure 3.5.* Dependence of energy yield versus planetary mill power.
*a* – kinetic data for the reaction $BaCO_3+WO_3=BaWO_4+CO_2$ at different frequencies, $s^{-1}$: 1 – 12.6; 2 – 11.03; 3 – 10.1; 4 – 9.2; 5 – 9.0; 6 – 8.9.
*b* – energy yield calculated on the basis of semi-transformation time ($\tau_{0.5}$).

## 3.4. Comparison of soft mechanochemical synthesis with other methods

Soft mechanochemical synthesis has many common features with other well-known and widely used methods, such as co-precipitation and hydrothermal synthesis. Really, co-precipitation provides components to be mixed at an atomic level. Just the same process is realized by mechanical activation of mixtures of hydroxides and hydrated oxides.

The formation of water (as a reaction product) and local heating under mechanical activation as a result of collisions, friction and plastic friction of solid particles are the conditions for the reactions to occur via a dissolved state (hydrothermal-like synthesis). Therefore, it is reasonable to consider some features of these methods taking into account a large number of works in this direction.

### 3.4.1. Specific features of the preparation of complex oxides from co-precipitated hydroxides

First, it should be noted that co-precipitation is realized only in the mixtures of hydroxides and hydrated oxides with solubility product below $10^{-10}$. Poorly soluble hydroxides are not very numerous (about 40) [9] but one can obtain a large number of compounds combining them with each other (or with several hydroxides) in different manners. In particular, the syntheses of complex oxides based on the binary mixtures of following hydroxides were described: $Mg(OH)_2$, $Cu(OH)_2$, $Zn(OH)_2$, $Ni(OH)_2$, $Co(OH)_2$, $Mn(OH)_2$, $Sn(OH)_2$, $Pb(OH)_2$, $Al(OH)_3$, $Fe(OH)_3$, $Cr(OH)_3$, $Mn(OH)_3$, $Sb(OH)_3$, $Bi(OH)_3$, $Si(OH)_4$, $Ti(OH)_4$, $Zr(OH)_4$, $Sb(OH)_5$, etc.

Definite regularities which govern the co-precipitation of hydroxides and the preparation of complex oxides by heating the co-precipitated complex hydroxides (CPH) are stated [10,11].

For example, the works cited in the review [11] showed that the mentioned processes are very sensitive to the deviations of the component ratio Me(II):Me(III) in solution from the stoichiometry corresponding to the composition of the final product $M^{II}M^{III}{}_2O_4$. When the composition deviates from stoichiometry, the co-precipitated hydroxides appear to be mechanical mixtures only. On contrary, when the composition exactly corresponds to the stoichiometry of the final product, chemical interaction occurs resulting in the formation of nano-sized X-ray amorphous product. This is evident by the data on the loss of chemical individuality of co-precipitated hydroxides. For example, in co-precipitated mixtures, like $Cu(OH)_2 – Me(OH)_3$, copper hydroxide becomes insoluble in ammonia and is not transformed into CuO under heating. For some oxides bound in this manner, the braking of dehydration is observed. X-ray amorphous product obtained by co-precipitation can be crystallized in the form of double hydroxide or even as a complex oxide.

The reasons of such behavior are connected with the differences in phase formation of solid product. Primary co-precipitation products appear to be soluble hydroxo-complexes of metals. Depending on rates of hydrolysis stages, the clusters of separate hydro-complexes or the clusters containing heteronuclear polymeric complexes are formed, then giving individual or complex hydroxides.

A very important fact was stressed by the authors [12-15] who investigated the crystallization of double hydroxides from co-precipitated mixtures. These authors discovered that different phase modifications are formed in the samples with rather close phase composition, depending on crystallization conditions. For example, in $Ni(OH)_2 - Al(OH)_3$ system at the composition range $(1.1–1.3)NiO \cdot Al_2O_3 \cdot (5–8) \cdot H_2O$, two different structures are formed with basal interplane distances of 0.80 and 0.86 nm. $Ni(OH)_2$ is known to possess brucite structure while $Al(OH)_3$ possesses bayerite structure. The authors demonstrated that the composition and the structure of final product depend on what structure (brucite or bayerite) serves as a matrix for its growing. The layer of the metal, which is most capable under given pH of solution to form hydroxide phase, plays a role of the packing that governs the precipitation. To achieve this, it is not necessary to isolate metal hydroxide as precipitate at first; it is sufficient to have a short-lived associate with the corresponding structural ordering in the liquid phase. These associates play the role of a nuclei programming the formation of layered packing.

A double hydroxide with the leading bayerite layer (d = 0.88 nm) is formed in $Ni(OH)_2 - Al(OH)_3$ system obtained by co-precipitation from acidic media enriched with aluminum, while the co-precipitation from alkaline media enriched with nickel results in the formation of the hydroxides with brucite layer in the packing (0.8 nm).

$Al^{3+}$ ions fill 2/3 of octahedral sites in the structure of bayerite layer composed of two layers of hydroxyl ions; the coupling of layers is provided so that each $OH^-$ group of one layer is located opposite to the $OH^-$ group of the next layer. $Me^{2+}$ cations occupy all the octahedral sites in brucite layer; each layer adjoins the other one due to $OH^-$ groups being located in triangular sites of adjacent layer. Therefore, the thickness of layer packing of brucite type is less than that of bayerite type.

At low molar ratio $NiO:Al_2O_3$ (<3.2), $Al(OH)_3$ is precipitated at first; it forms the structure of final product. For ideal packing of layers, the composition should correspond to $3NiO \cdot Al_2O_3 \cdot 6H_2O$. At a constant 1:1 ratio of these layers, variable composition of the phase with d = 0.88 nm is the evidence that bayerite layers are packed ideally in its structure while brucite ones, $Ni_3(OH)_6$, are discrete, i.e. possess "building" defects.

The formation conditions, the composition ($NiO:Al_2O_3$ = 4–14) and the characteristic basal distance d = 0.8 nm of the phase agree with another model of layer packing: a continuous brucite layer is a leading layer while bayerite layer possesses discrete structure, i.e., it is not tightly packed with $Al^{3+}$ cations.

According to the crystallographic principle concerning maximum filling of space in lattice [16], "building" defects of the discrete layer should be filled. Water molecules fill them during the synthesis from water-salt systems. For $[Ni_3(OH)_6]_{2\infty}$ leading layer, 9÷21 water molecules enter the double hydroxide structure, while for $[Al_2(OH)_6]_{2\infty}$, the amount of water molecules is equal to 7÷9.

In the intermediate case, the ions of divalent metal do not form their own structural layer but are localized in the vicinity of vacant octahedrons $[Al_2(OH)_6]_{2\infty}$ using $OH^-$ groups of this layer, as well as $OH^-$ groups and $H_2O$ molecules of the freshly formed third network of layered packing. In this case, interplanar distance is equal to 0.76 nm. Hence, the crystal chemical formula for this compound is $[Me(OH)_2{\cdot}H_2O][Al_2(OH)_6]$. It was obtained in $Mg(OH)_2$ – $Al(OH)_3$ system. The decomposition of such compounds at moderate temperatures (about 800°C) can result in the formation of stoichiometric spinels, which was demonstrated in [15].

Thermal decomposition (dehydration) of air-dried X-ray amorphous CPH proceeds in different ways. Similar to the decomposition of X-ray amorphous hydroxides, it occurs gradually at a wide temperature range, while the total mass loss at isothermal dehydration gradually increases at elevated temperatures.

The primary product of dehydration, which is an oxide phase with low content of chemically bound water, is X-ray amorphous as a rule, and under thermal treatment crystallizes with the formation of nanocrystals of complex oxide. If the composition differs from stoichiometry, then the phase of the excess oxide is also precipitated.

Mössbauer spectroscopy was used to establish the sequence of stages in the processes of phase formation which occur in the systems based on iron hydroxide: $Fe(OH)_3$ – $Cu(OH)_2$, $Fe(OH)_3$ – $Ni(OH)_2$, $Fe(OH)_3$ – $Si(OH)_4$, etc. [11]. It was stated that similar iron hydroxide clusters with local cubic symmetry ordering were formed in the systems $Fe(OH)_3$ – $Cu(OH)_2$, $Fe(OH)_3$ – $Ni(OH)_2$, as well as in separately precipitated $Fe(OH)_3$. In both cases, Mössbauer spectrum is gradually transformed from doublet into sextet simultaneously with the removal of water. At temperature about 400°C, the profound rearrangement of the structure of primary dehydration product starts, which results, according to Mössbauer spectra, in an appearance either of a new disordered $\alpha$-$Fe_2O_3$ phase when $Fe(OH)_3$ being heated, or ferrite if $Fe(OH)_3$ – $Me(OH)_2$ is heated. In the latter case, a temperature of 500°C is enough to obtain pure ferrite phase via dehydration.

For $Fe(OH)_3$ - $Si(OH)_4$ system, as it follows from Mössbauer spectra, Fe–OH–Si or Fe–O–Si groups are formed directly avoiding the above-mentioned stage of the formation of individual clusters.

High rate of the formation of complex oxides from co-precipitated hydroxides or mixtures of hydroxides can be explained by the conjugation of dehydration and synthesis reactions, stated by the authors [17-18]. To provide conjugation, close or overlapping temperature ranges of dehydration and interaction are needed, as well as the possibility to form cation vacancies in the lattice of acceptor phase, with oxygen coordination corresponding to the preferable coordination for cations from donor phase.

This is realized at the dehydration of $Al(OH)_3$ – $Cr(OH)_3$ mixture [17]. Depending on the nature of aluminum hydroxide, cation vacancies are distributed in different manners at heating. For example, $\eta$-$Al_2O_3$ is formed under dehydration of bayerite Al(OOH). It possesses cation vacancies in tetrahedral sites $\square_{2.67}Al_{5.33}[Al_{16}]O_{32}$, while octahedral sites are completely occupied. Dehydration of boemite $\alpha$-AlOOH brings to $\alpha$-$Al_2O_3$ with partially vacant octahedral sites and tetragonal distortion of the unit cell $Al_8[\square_{2.67}Al_{13.13}]O_{32}$ in which all the tetrahedral sites are occupied. Both oxides, $\eta$-$Al_2O_3$ and $\alpha$-$Al_2O_3$, possess practically identical structures and give similar diffraction patterns. They can be distinguished by comparing intensity ratios $I_{311}/I_{222} = k$. For $\alpha$-$Al_2O_3$, k is close to 1, while for $\eta$-$Al_2O_3$ k is close to 2.

Heating of $\alpha$-AlOOH and $\alpha$-CrOOH mixtures at 700°C for 6 h results in the formation of aluminium chromium spinel with a structure similar to $\eta$-$Al_2O_3$, in which Cr(III) ions are present only in octahedral coordination. A maximum amount of Cr(III) that can enter spinel lattice coincides with a maximum number of sites in $\alpha$-$Al_2O_3$ (which is equal to 2.67). Since dehydration temperatures of $\alpha$-AlOOH and $\alpha$-CrOOH are close to each other, the occupation of octa-sites in $\alpha$-AlOOH occurs at the very moment of their formation, i.e., when the dehydration of $\alpha$-AlOOH and lattice rearrangement occur. The occupation of octa-sites of Cr(III) is accompanied by the formation of an equivalent number of vacant tetrahedral sites (which is due to the necessity to conserve lattice electroneutrality) and by a gradual transformation of the spinel lattice into the structural type of $\eta$-$Al_2O_3$. After all octahedral sites are occupied, the excessive Cr(III) ions are bound to give an individual $\alpha$-$Cr_2O_3$ form.

For $\alpha$-AlOOH – $Zn(OH)_2$ and $\eta$-AlOOH – $Zn(OH)_2$ mixtures, in which the tetrahedral coordination is preferable for zinc ions, spinel is formed only in the case of $\eta$-AlOOH [18]. However, spinel as prepared is nonstoichiometric; it corresponds to the formula $Zn_{0.29}Al_{2.47}O_4$ arising from the number of tetrahedral positions in $\eta$-$Al_2O_3$. No conditions for conjugation exist in $\alpha$-AlOOH – $Zn(OH)_2$ mixture, therefore, X-ray diffraction patterns exhibit only the reflections of well crystallized boemite and ZnO. The formation of spinel starts only at high temperatures (above 850°C) according to diffusion mechanisms of solid-phase reactions.

The regularities and mechanisms described above are, probably, realized in the course of phase formation in the mixtures of hydroxides and hydrated oxides under mechanical activation.

### 3.4.2. Peculiarities of hydrothermal synthesis

Hydrothermal synthesis, developed in the second half of the XIX century by French mineralogists for modeling natural processes, is now widely used to prepare many compounds of practical importance [19].

This method is based on the ability of water to dissolve the substances at increased temperature and pressure. The use of water as a solvent is based on self-ionization: $2H_2O \rightarrow H_3O^+ + OH^-$.

Ionization degree increases with temperature rise. When some limiting amount of a substance dissolved in solution is exceeded (limiting supersaturation), its spontaneous crystallization starts. This process can be governed by changing of the temperature (and pressure, correspondingly) as it is carried out in a closed vessel.

Substantial differences between solid-phase reactions and hydrothermal synthesis reactions have been stated in numerous investigations. In solid-phase processes, the sequence of intermediate products formation does not depend on reagents ratio in the initial mixture, and the excess product appears to be a compound with the highest crystallization temperature. On contrary, for the formation of a definite product by hydrothermal synthesis, the initial mixture should contain reagents at an exact stoichiometric ratio [19,20]. In solid-phase reactions, the interaction rate is determined by the rate of diffusion processes, while in hydrothermal processes the determining factor is the rate of dissolution of the initial products in the water. Water simplifies diffusion transport of particles in the system; the formation of nuclei and crystal growth occur faster than in solids.

Hydrothermal synthesis reactions are accelerated substantially in the solutions of NaCl, KCl, $NH_4Cl$, etc. The catalytic action of salts is connected with the increase of dissolution rate of initial compounds in the presence of salts. The treatment of mixtures of co-precipitated substances and gels under hydrothermal conditions provides a substantial acceleration of their crystallization.

A large number of investigations on the synthesis of inorganic compounds by hydrothermal method was carried out. At present, the list of compounds that are synthesized under hydrothermal conditions includes aluminates, ferrites, chromates, titanates, zirconates, stannates, antimonates, etc. The works dealing with the synthesis of silicates are most numerous [19-26].

Aluminates are formed at temperature above 500°C and pressure of 100 MPa in pure water from poorly soluble oxides according to the reaction $MeO + Al_2O_3 = MeAl_2O_4$. In order to achieve a 100 % yield, it is necessary for the molar ratio of the components to be close to 1:1. Aluminates are formed as fine crystals (smaller than 0.1 nm) without being transported into the cool zone of the autoclave. Aqueous solutions of NaOH are the most favourable ones for this synthesis. The method was used to prepare the aluminates of Be, Ca, Zn, Fe, Co, Ni [21-23].

For the ratio $MeO/Al_2O_3 > 2:1$, hydrogarnets of $3MeO \cdot Al_2O_3 \cdot 6H_2O$ type are usually formed. In pure water, their synthesis occurs at temperature of 200-400°C and pressure of 50-80 MPa. A 100 % yield is observed when the reagents ratio $MeO/Al_2O_3$ is equal to 3:1. In aqueous solutions of NaOH, the stoichiometry can deviate towards MeO excess.

In hydrothermal sŷnthesis, $3CaO \cdot Al_2O_3 \cdot 6H_2O$ is formed as transparent crystals of tetragontrioctahedrons, less frequently as cubes and rhombododecahedrons. The synthesis takes place under large excess of CaO with respect to $Al_2O_3$ (more than 5:1) within temperature range 300-400°C. Tri-strontium aluminate hexahydrate crystals are formed in a similar way [23].

The interaction of titanium dioxide with the oxides of mono- and divalent metals occurs under hydrothermal conditions also. In alkaline solutions with 5-70 mass.% of NaOH, a series of sodium titanates is crystallized: $Na_2O \cdot 3TiO_3$, $Na_2O \cdot 4TiO_2$, $Na_2O \cdot 5TiO_2$, and $Na_2O \cdot 6TiO_2$ [24]. Only one representative of titanates, namely, $Na_2O \cdot TiO_2$, is formed in aqueous solutions of KOH.

Metatitanates of divalent metals Ca, Sr, Ra, Cd and Pb were obtained in aqueous solutions of NaOH and KF [25]. The synthesis is carried out using $Me(OH)_2 + TiO_2 + H_2O$ mixture at 380-450°C and pressure 30-50 MPa. The isometric crystals are formed. Since the formation of titanates occurs in the presence of an alkaline metal, the synthesis is carried out in low concentrated NaOH solutions (~10 % mass.) The synthesis of these compounds proceeds better if metatitanium acid, $H_2TiO_3$, is used instead of titanium dioxide [25].

100% yield of lead titanate was obtained in aqueous solutions of KF (10 % mass.) at temperature of 450-700°C and pressure of 70-300 MPa [25]. At the ratio $PbO/TiO_2 = 1/1 \div 2/1$, dark-green crystals with 1-2 nm size were obtained.

Fine crystalline quartz or amorphous substances (gel, quartz glass, etc.) are used as one of initial components for the synthesis [19]. Depending on temperature, pressure, pH of the medium and the presence of salts, silica can exist in solution both as simple ions or molecules, and as more complicated polymer particles. Under normal conditions silica passes into solution in monomer form, as silicic acid $Si(OH)_4$; at large pH, silicate ions $SiO_3^{2-}$ are formed. Monosilicic acid is a very weak acid; however, at increased temperature its dissociation constant increases substantially. The amount of monomer form also increases with temperature. The dissolution of $SiO_2$ is due to hydration as well as to depolymerization.

The rate of silica dissolution is proportional to its surface area. Silicic acid being passed into solution undergoes polymerization giving water and the molecules of polysilicic acids of variable complexity. At pH within 5-7, silicic acid polymerizes to form colloid solutions, followed by their coagulation. If NaOH is added, $SiO_2$ amount in the aqueous solution increases sharply. An increase of sodium concentration from 0 to 5.8 mass.% causes an increase of the solubility of quartz from 0.27 to 14.65 mass.%.

With silica taken in large excess, only two silicates are crystallized at 450°C: $Na_2O{\cdot}2SiO_2$ and $Na_2O{\cdot}SiO_2$. The former is crystallized when NaOH concentration is above 8 % and silica content is about 40 %; the latter is crystallized when NaOH content is 20–30 %. In the solutions with low NaOH concentration, only $Na_2O{\cdot}3SiO_2{\cdot}0.5H_2O$ is formed.

Reactions that take place under hydrothermal conditions in $CaO - SiO_2 - H_2O$ system are complicative; the number of silicates synthesized in this system approaches 20 [26]. The state of calcium hydroxide has a substantial effect on final products. If amorphous hydroxide is used, the reaction runs faster, but at the same time the probability of metastable phases formation increases. Detailed investigations that were carried out by the authors [26] allowed to state the preferable regions of the synthesis of various calcium compounds depending on $CaO/SiO_2$ ratios and hydrothermal treatment temperature (450-850°C).

The solubility of calcium hydroxide is low (below 0.1 g/l at 300°C) and decreases with temperature increasing. Therefore, aqueous solutions are always saturated with respect to portlandite, while silicon concentration is variable (increases with temperature).

If the reaction follows the first route, the interaction of monosilicic acid with calcium hydroxide molecules appears to be a direct reason of polymerization and precipitation of hydrated calcium silicates. The scheme of the reaction can be written as follows [20]:

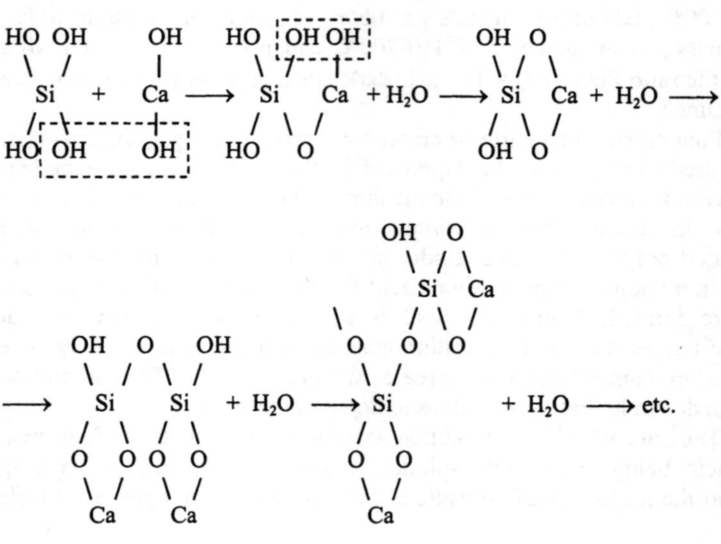

Silicium oxygen radical passes from discrete ortho- and diorthogroups (separately or jointly) to an infinite chain, then to infinite ribbon $[Si_6O_{17}]_\infty$. Xonothlite is the final product of the phase sequence. Three-dimensional bonds composed of $CaO_6$ polyhedrons in γ-dellaite, $Ca_6[SiO_4][Si_2O_7](OH)_2$, transform to discrete walls composed of $CaO_6$ polyhedrons in xonothlite, $Ca_6[Si_6O_{17}](OH)_2$, i.e. the decreasing of the condensation degree occurs.

Calcium silicates formed according to the topochemical mechanism should contain portlandite blocks (layers) in their structure. Experiment shows that tri-calcium silicate is formed by the interaction of portlandite with unsaturated silica solutions. With the increasing of silica concentration, the following compositions are crystallized: $Ca_6[SiO_4]_2{\cdot}H_2O$, $Ca_5[SiO_4]_2(OH)_2$, $Ca_8[Si_5O_{18}]$, $Ca_6[Si_4O_{14}]$. This means that kilchoanite, $Ca_6(Si_4O_{14})_4$, will be the final member since it is impossible to increase silica concentration without disturbing the crystallization mechanism.

In [27-29], hydrothermal, mechanochemical and solid-phase syntheses of calcium silicate from anhydrous and hydrated oxides were compared. Initial components were taken at Ca/Si ratio equal to 0.8:1.0:1.2. According to X-ray analysis, the interaction in the mixtures of anhydrous oxides under mechanical activation is not completed. The product being formed is X-ray amorphous. When heated at 600-800°C, it is crystallized in the form of α'-$Ca_2SiO_4$ (Fig. 3.6*a*). At higher temperatures, $Ca_3Si_2O_7$ is formed; β-$CaSiO_3$ is crystallized at 860°C. The observed sequence of stages is similar to those observed in solid-phase synthesis of wollastonite. The amount of β-$CaSiO_3$ at 900°C does not exhibit any substantial dependence on the initial fractions of the reagents.

The interaction of $Ca(OH)_2$ with $SiO_2{\cdot}0.5H_2O$ is more intensive. The products formed in this mixture are shown to be X-ray amorphous (Fig. 3.6*b*). They contain some amount of bound water. TG analysis showed that different molar ratio of the components leads to different products: $5CaO{\cdot}6SiO_2{\cdot}5H_2O$, $6CaO{\cdot}6SiO_2{\cdot}7H_2O$ and $5CaO{\cdot}6SiO_2{\cdot}5H_2O$, correspondingly, in the mixtures 0.8, 1.0 and 1.2. The products in 0.8 and 1.0 mixtures remain weakly crystallized until 820-850°C, then transforming into β-$CaSiO_3$, while the crystallization of the product in 1.2 mixture begins at 600°C. Thus, the interaction of $Ca(OH)_2$ with the hydrated silica is strongly dependent on molar ration of reagents. This fact is indicative of the possible role of liquid-phase processes similar to hydrothermal synthesis.

After thermal treatment at temperatures above 850°C, all the hydrosilicates decompose to give wollastonite β-$CaSiO_3$.

It was stated that the adding of free water to the mixture of anhydrous oxides influences on the kinetics of their mechanochemical interaction. It has an extremal dependence: the degree of CaO bonding into silicates is maximal at 2-5 mass. %.

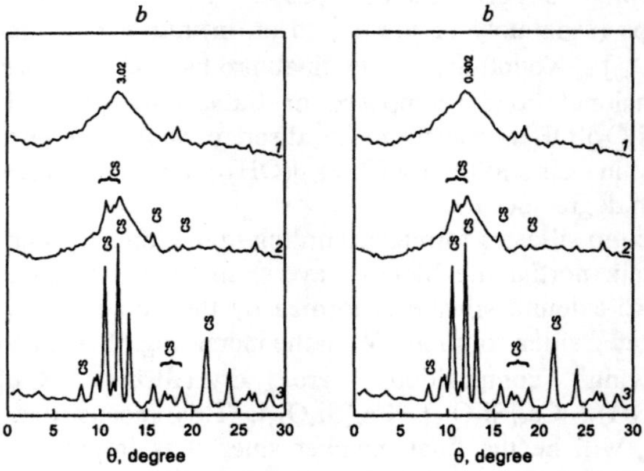


*Figure 3.6.* X-ray profiles of the products of mechanochemical reactions between CaO and anhydrous $SiO_2$ (a); $Ca(OH)_2$ and $SiO_2aq$. (b):
1 – after activation for 10 min, 2 - 600°C, 3 - 900°C.
CS – $CaSiO_3$, $C_2S$ – $Ca_2SiO_4$, $C_3S_2$ – $Ca_3Si_2O_7$.

The scheme of these processes can be written as follows:

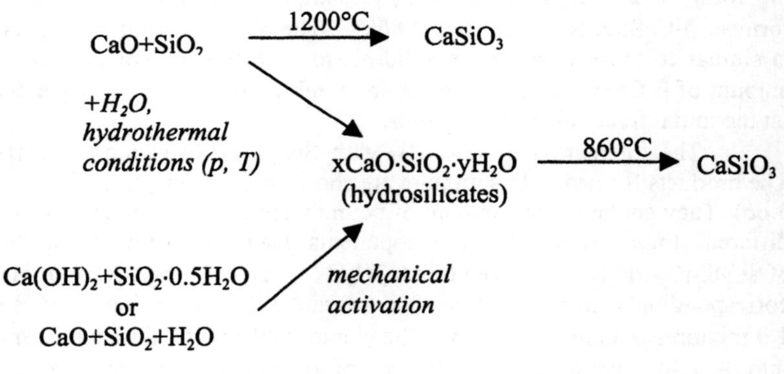


Thus, based on above-considered data one can conclude that hydrothermal and mechanochemical synthesis brings to similar products – hydrated calcium silicates.

### *References*


1. Kireev V.A. Acid-base properties of oxides. Zhurn. Fis. Khim. 1964; 38: 1881-92.
2. *Thermal Constants of Substanses.* In 10 volumes. Ed. Gluschko V. P. , Moskva: VINITI, 1966-1988.
3. Viting L.M., Avvakumov E.G. About correlation between Gibbs energies and average orbital electronegativities of complex oxides. Zhurn. Fis. Khim. 2001; 75: 1120-22.
4. Avvakumov E.G., Kosova N.V. Fast propagating solid-state mechanochemical reactions. Sibir. Khim. Zhurn. 1991; 5: 62-66.
5. Avvakumov Evgenii, *Mechanical Methods of Activation of Chemical Processes.* Novosibirsk:Nauka, 1986.
6. Khodakov, Genrich, *Physics of Grinding.* Moskva: Nauka, 1972.
7. Cocco G., Delogu F., Schiffini L. Toward a quantitative understanding of the mechanical alloing process. J. Mater. Synth. and Proc. (to be published).
8. Butyagin P.Yu., Pavlichev J.K. Energetic yields of mechanochemical reactions. Reactivity of Solids 1986; 1, 345-61.
9. Kumok V.N., Kuleschova O.M., Karabin L.A. *Solubility Product.* Novosibirsk: Nauka, 1983.
10. Dzisko V.A., Karnaukhov A.P., Tarasova D.V. *Physico-Chemical Bases of Synthesis of Oxide Catalysators.* Novosibirsk: Nauka, 1978.
11. Sviridov V.V., Braniczkii G.A. Nanostructural hydroxides, oxides and metal-oxides systems. In *Chemical Problems of Preparation of New Materials and Technologies.* Ed. Sviridov V.V., Minsk; Bel. Gos. Universitet, 1998.
12. Tomilov N.P., Berger A.S. About double hydroxides of nickel and aluminium of layered structure with leading bayerite or brucite layers. I. Phase variety of double hydroxides of nickel and aluminium, prepared by different methods. Izvestiya SO AN SSSR, ser. khim.nauk 1984; 1: 40-48.
13. Tomilov N.P., Ivashina V.S., Berger A.S. About double hydroxides of nickel and aluminium of layered structure with leading bayerite or brucite layers. II. Double hydroxides of nickel and aluminium at sorption equilibrium in the system sodium hydroalumocarbonate - nickel nitrate solution. Izvestiya SO AN SSSR, ser. khim. nauk 1984; 1: 48-52.
14. Morozkova V.E., Tomilov N.P., Berger A.S. About double hydroxides nickel and aluminium of layered structure. III. Sorption of water vapour by phase variety of nickel and aluminium hydroxides. Izvestiya SO AN SSSR, ser. chem. nauk, 1984; 6: 50-57.
15. Tomilov N.P., Deviatkina E.T. Synthesis of $MgAl_2O_4$ from coprecipitated hydroxides. Neorgan. Materiali 1990; 26: 2556-61.
16. Vainsthein, Boris, Fridkin Vladimir, Indenbom Vladimir. *Modern Crystallography. V. 2. Structure of Crystals.* Moskva: Nauka, 1979.
17. Krivoruchko O.P., Plyasova L.M., Zolotovskii B.P., et al. Investigations of formation mechanism and cation ditributions in $Al^{III}$-$Cr^{III}$ spinels. Zhurn. Neorgan. Khim. 1979; 24: 2913-18.
18. Taraban E.A., Krivoruchko O.P., Buyanov R.A. et al. Dehydration reactions and solid-state interactions in thermal treatment of mechanical mixtures of crystal hydroxides $Al^{III}$ with hydroxides $Zn^{II}$. Zhurn. Neorgan. Khim. 1988; 33: 1824-28.
19. Litvin, Boris, and Popolitov, Vladimir. *Hydrothermal Synthesis of Inorganic Compounds.* Moskva: Nauka, 1984.
20. Kalinin, Dimitrii, *Mechanism and Kinetic of Hydrothermal Reactions of Silicates Formation.* Novosibirsk: Nauka, 1973.
21. Demianecz L.N. Hydrothermal synthesis of transition metal oxides. *In: Hydrothermal Synthesis and Monocrystal Growth.* Moskva: Nauka, 1982: 4-26.
22. Crowley M.S. Effect of starting materials on phase relations in the system $CaO$-$Al_2O_3$-$H_2O$. J. Am. Ceram. Soc. 1964; 47: 144-48.

23. Ponomarev V.J., Litvin B.N., Belov N.V., Hydrothermal synthesis of calcium hydroaluminates and their X-ray analysis. Isvestiya AN , Neorgan. Materiali 1970; 6: 1657-59.
24. Hirano S., Ismail M.G., Soniya S. Crystal growth of sodium titanate bronze compounds under hydrothermal conditions. Mater. Res. Bull. 1998; 11: 1013-30.
25. Barsukova M.L., Kuzneczov V.A., Malinovskay E.K. Crystalization of bivalent metals titanates under hydrothermal conditions. Kristallografiya 1972; 17: 1268-70.
26. Ilukhin V.V., Kuzneczov V.A., Lobachev A.N., Bakschytov V.C. *Calcium Hydrosilicates*. Moskva: Nauka, 1979.
27. Komljenovic M.M., Radakovic A., Zivanovic B.M. et al. An investigation of solid- and liquid-state reactions in $CaO\text{-}SiO_2$ system. Science of Sintering 1994; 26: 185-93.
28. Boldyrev V.V., Khabibullin A.Kh., Kosova N.V., Avvakumov E.G. Hydrothermal reactions under mechanochemical treatment. J. Mater. Synthesis Proc. 1996; 4: 377-81.
29. Kosova N.V, Khabibullin A.Kh., Boldyrev V.V. Hydrothermal reactions under mechanochemical treating. Solid State Ionics 1997; 101-103: 53-58.

# Chapter 4

# APPARATUS FOR MECHANOCHEMICAL REACTIONS

Mechanochemical reactions are carried out by means of mechanical activation in high energetic apparatus. Grinding devices with high load rates are used for this purpose, such as planetary, vibratory centrifugal, shock-reflective and ring roll mills, attritors, etc.

## 4.1. Planetary mills

Relatively new and promising high energetic planetary and differential centrifugal mills were developed by Bushuev, Golosov and other authors [1-6]. High rate of grinding and activation in these mills is provided by centrifugal forces arising at the rotation of each jar around its own axis and their common axis. These centrifugal forces exceed the gravity force by a factor of several tens, thus, providing the energy of these mills by 2-3 orders of magnitude higher than that of common ball mills. Powder dispersion achieved in a centrifugal mill after 2 min treatment is the same as that achieved in usual ball mill after the treatment for 10–12 h. The most of results reported in the present book were obtained mainly with planetary mills of periodic action that were improved and adapted to carry out mechanochemical reactions in a closed system with external cooling.

In new laboratory planetary mills, the jars rotate due to frictional clutch with the cylindrical housing of the mill. A scheme and the design of the apparatus are shown in Fig. 4.1 [5]. The apparatus consists of the housing 1, carrier 2, jars 3, guiding rim 4, sheave 5, channels in the center and at the periphery of the carrier for cooling water. The diameter of the jars is 70 mm, the working volume is 150 $cm^3$, ball mass is 200 g, the mass of the substance under treatment is 10 g.

The device operates as follows. At first, water is admitted from water pipeline through the tube 6 and channels 7 into the pockets 8. When the motor (not shown in the scheme) is switched on, the carrier starts to rotate; the rings that are fixed on it make jars rotate. The surfaces of jars and rings are separated from each other by water layer, like in hydrostatic bearings: the jars operate as a spindle while the ring acts as hydrostatic bearing. Under the action of centrifugal forces, the jars are pressed to the guiding rim 4 and due to frictional clutch with its surface rotate around their axis. Thus, the jars take part in two kinds of rotation: around their common and own axes. This technical solution excludes the necessity to use bearings in the most loaded part of the mill (the drive of jars) and increases operational lifetime of the apparatus. Intensive cooling of the jars by water decreases the contamination of the product with the material of grinding

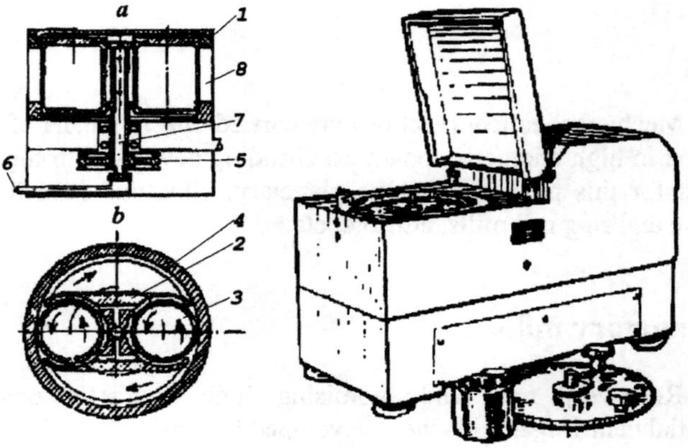


*Figure 4.1.* The scheme of operation and design of the AGO-2 planetary mill [5]. *a* – view from side, *b* – view from above.

bodies as well as increases the degree of product activation since the annealing of defects is excluded.

In order to characterize the operational regime of the planetary mill, k and m parameters were introduced. $k = \omega_2/\omega_1$ (here $\omega_2$ is a frequency of rotation around the planetary axis and $\omega_1$ is a frequency of rotation around the common axis), and $m=R_2/R_1$ ($R_1$ and $R_2$ are the radii of common and planetary rotation).

The value of k can be either positive or negative. This depends on the direction of vectors of relative and planetary rotation of the jars. When these directions are the same, k is positive, but if the vectors are directed opposite to each other, k is negative. The most efficient regimes of the planetary mill are within the range when k = -1.

According to [4], theoretical power of the planetary mill can be calculated according to the equation:

$$W_{plan.} = (W_{ball} \cdot 4\pi^2 R_1 \omega^2 / 981) \cdot n_1 / n_2 \qquad (4.1)$$

here $W_{ball}$ is equal to the power of the ball mill:

$$W_{ball} = 6.1\, P \sqrt{D}, \qquad (4.2)$$

P is a mass of balls, kg; D is a jar diameter, m; $\omega_1$ is a frequency of rotation (of the rotor), $s^{-1}$; $n_1$ is a critical rotation frequency of the planetary mill which is equal to

$$n_1 = 0.8 \sqrt{R_1/R_2}; \quad (4.3)$$

$n_2$- critical number of revolution in ball mill which is equal to

$$n_2 = 32 \sqrt{D}. \quad (4.4)$$

Energy parameters of laboratory planetary mill were determined experimentally in [7,8]. The power was determined by means of calorimetry. Water flow temperature was measured at the inlet ($T_1$) and at the outlet ($T_2$); the calculation was performed according to the equation

$$Q = mc(T_2 - T_1), \quad (4.5)$$

where Q is the power scattered by a jar as a heat, m is a mass, C is a heat capacity of water.

A blank experiment was carried out with balls being replaced by immobile plate of equal mass. This contribution was subtracted from the total energy consumed. The results are shown in Table 4.1.

*Table 4.1.* The power of the laboratory planetary mill determined experimentally (balls mass 100 g, the substance to be treated: 10 g) [8].

| The frequency of total rotation, rpm | Power, W | |
|---|---|---|
| | Ball diameter: 5 mm | Ball diameter: 9 mm |
| 602 | 191 | 104 |
| 858 | 271 | 204 |
| 996 | 602 | 389 |

As one can see, energy consumption by the present mill is equal to 10÷60 W/g.

Besides total energy consumption for mechanochemical reactions, the level of local parameters of temperature and pressure arising in the mill is also important. It is very difficult to measure these parameters directly during activation. Therefore, an indirect method based on P and T determination of phase transitions under mechanical activation of the substances with known P-T phase diagrams was used [9]. The results are shown in Table 4.2.

*Table 4.2.* Phase transitions observed in a planetary mill (the frequency of total rotation: 856 rpm) [9].

| Phase transition | P, GPa | T, K | Result |
|---|---|---|---|
| β-$PbO_2$→ →α-$PbO_2$ | 1.3 | 0 – 673 | + |
| H-$Nb_2O_5$→ →T-$Nb_2O_5$ | 1.1 – 1.5 | 400 – 500 | + |
| $Pb_3O_4$→PbO+ +$Pb_2O_3$ | 1.8 – 3.0 | 473 – 673 | + |
| T-$Nb_2O_5$→ →B-$Nb_2O_5$ | 3.0 | 773 | - |
| $TiO_2$→$TiO_2$ (high pressure) | 4.0 | - | - |

It follows that in the mill, local temperatures arise up to 600–800K and pressure - within 2–3 GPa. Other authors confirmed these data

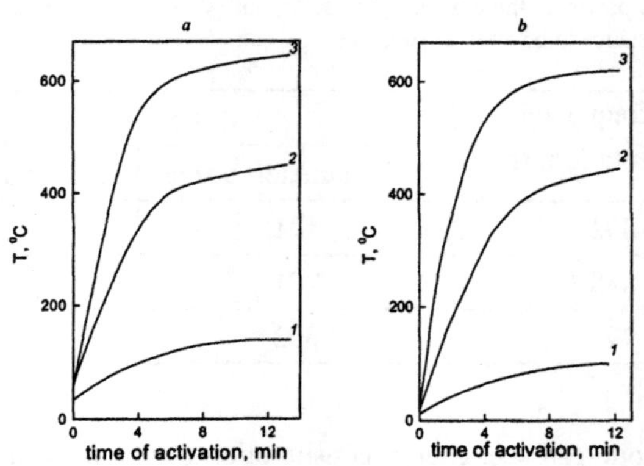


*Figure 4.2.* Changes of the temperature of balls in the planetary mill, depending on activation time, rotation frequency, and ball size [8].
*a* – rotation frequency, $s^{-1}$: 1 – 10.2, 2 – 14.3, 3 – 16.6,
*b* – ball diameter, mm: 1 – 3, 2 – 5, 3 – 9.

experimentally. For example, in [8] the ball temperature was determined from the calorimetric data. The authors calculated ball temperature according to heat balance equation having determined a total heat of the jar and balls. The data for different rotation frequencies and ball diameters are shown in Fig. 4.2. With increasing the ball size and the frequency of total rotation, temperature increases and reaches a level of about 873 K. High temperature of large balls is due to the generation of heat at inelastic deformation of balls during their collisions with a wall and due to worse heat exchange conditions in this case.

In [10], a comparison of the power of different mills at optimal regimes of their operation was carried out. Some of the results are shown in Table 4.3.

*Table 4.3.* Comparison on the power of different mechanochemical activators [10].

| Mill type | Frequency, $s^{-1}$ | Ball diameter, mm | Ball mass, g | Mass of powder, G | Power per 1 g of powder, W/g |
|---|---|---|---|---|---|
| Planetary, AGO-2 | 11.4 | 9<br>5 | 100 | 4 | 10<br>7 |
| Planetary, Pulveri-zette 7 | 4 | 9<br>7 | 53 | 2<br>4 | 0.3<br>0.09 |
| Vibratory, M-35 | 50 | 9 | 107 | 10 | 12 |
| SPEC-8000 | - | 6 | 307 | 2 | 9 |

As one can see, among the mills used for mechanochemical investigations, AGO-2 planetary mill exhibits better characteristics.

A middle-scale mill AGO-3 was constructed for operation in periodic regime [6]. A principle of hydrostatic bearing in AGO-3 mill is the same as in AGO-2 mill. The jars rotate around the common and own axes due to frictional engagement with the surface of the cylindrical guide surface. The volume of a jar (their number in the carrier being three) is 2 $dm^3$. Total ball mass is 2 kg; the mass of substance under treatment is 300 g per one jar. The amount of product that can be prepared within one working day is equal to 7 ÷10 kg.

## 4.2. Vibratory-centrifugal mills

Vibratory-centrifugal mills, developed by Denisov with coauthors [11] and marked as 5G, 10G, 30G, 50G, could be successfully used for continuous operation to carry out mechanochemical syntheses under industrial conditions.

Fig. 4.3 shows the design of this vibratory mill. It is constructed with horizontal position of the milling jars with the tube form. They are rigidly fixed on carriers and move together with them along the circle trajectory in a plane perpendicular to the axes of jars; the jars do not rotate along their own axes; their orientation in the vertical plane remains unchanged.

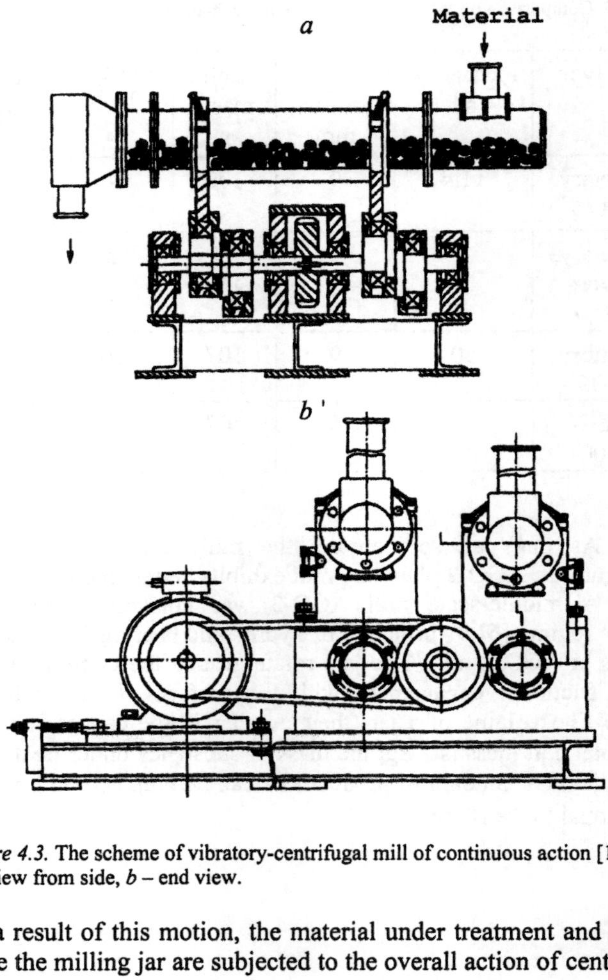


*Figure 4.3.* The scheme of vibratory-centrifugal mill of continuous action [11]. *a* – view from side, *b* – end view.

As a result of this motion, the material under treatment and milling bodies inside the milling jar are subjected to the overall action of centrifugal, gravity and Coriolis forces. This creates inside the jar the conditions providing the material to be subjected to intensive action of vibration shock, attrition and crushing loads that results in grinding and activation of the treated material.

The main working elements of the mill are eccentric crankshafts with the pairs of eccentrics on them. The eccentrics in a pair are counter to each other, i.e., their eccentricities lie strictly in one plane and are opposite directed. At synchronous rotation of crankshafts, grinding blocks connected to opposed eccentrics move around circles, their motions being shifted in phase by $180^{\circ}$ with respect to each other. At every moment they are in static equilibrium; thus, the mill does not require balancing adjustment, except the following: the weight of milling bodies in each jar as well as the weight of treated material should be equal per unit time.

The rotation is passed to the drive shaft of the motor through a V-belt transmission. The drive shaft rotates crankshafts synchronously and transmits the motion to milling blocks. The productivity of the mill depends on the size of jars, the frequency of shaft rotation, rate of material admittance through the jars, etc. Centrifugal mills providing the productivity of 5÷50 kg per hour have been constructed.

## 4.3. Ring roll mills

The ring roll mills MICROS, constructed and produced by Nara Machinery Co., LTD, Tokyo appeared to be high-effective for soft mechanochemical reactions to occur. The grinding of the particles is realized by compression and shearing forces generated by the rotating grinding rings. The scheme of the mill and the principles of its work are presented in Fig. 4.4.

The main body of the MICROS comprises of a casing, revolving main shaft in the casing and several sub-shafts interlocked with the rotation of the main shaft. Each sub-shaft has a number of rings shaped grinding media. The size of the ring as a grinding body is different depending upon models. The outside diameters are of 25÷45 mm and thickness of several mm.

The equipment is designed in such a way that each ring moves freely, allowing a small space between the outside diameter of the sub-shaft and the inner diameter of ring. The ring moves radial as much as the space allows by the centrifugal force created by the rotating main shaft and keeps turning by being pressed toward the inside wall of the casing. At this moment, the ring rotates itself around the sub-shaft by the friction force with wall surface. This means that while the ring revolves in the casing, the latter also rotates.

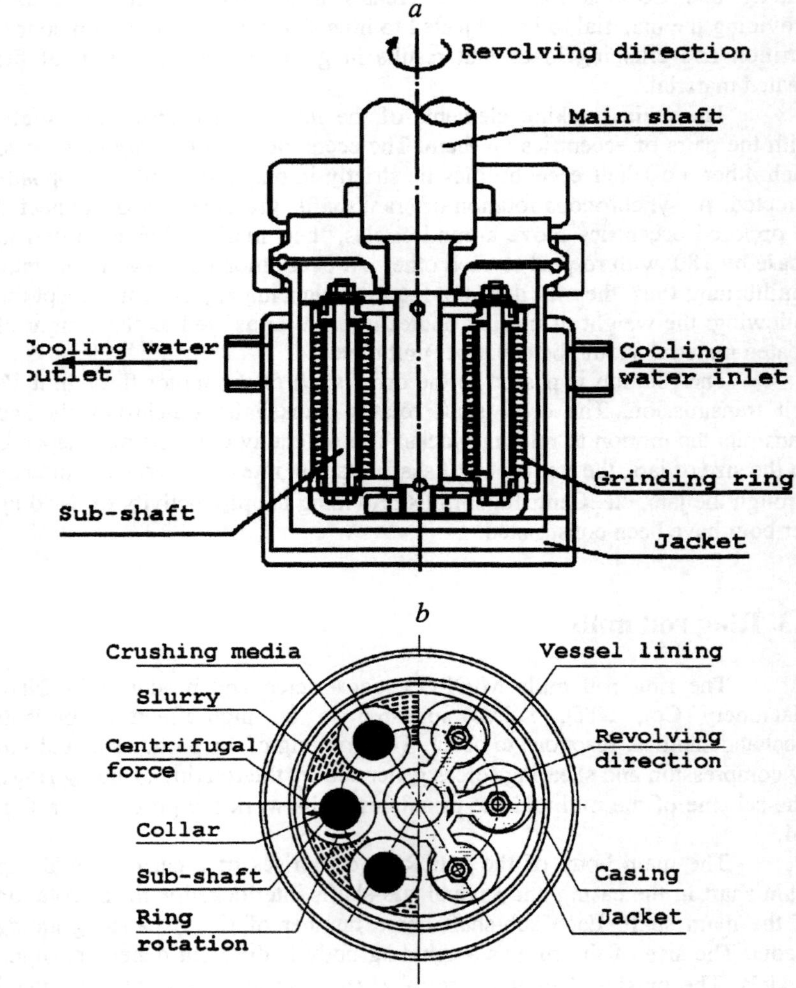


*Figure 4.4.* The scheme of the operation of MICROS ring roll mill [12].
*a* – view from side, *b* - view from above.

The particles are held between the revolving ring and the wall surface, and are subjected to compressing force created by the centrifugal force of the ring and the friction created by the rotation ring. This process generates pulverization and dispersion of the particles.

Since the rings are allowed to move freely, an independent crushing effect is given to the particles even when the fed material contains small and large particles. The casing's structure is easily removable along with the rings and other components. As a result, inspection and maintenance are very easy.

The casing is equipped with a jacket which is used to control heating during milling by sending chilled water into the jacket. The construction material available for the casing's inner wall and the rings are as follows: stainless steel, ceramics or super hard materials etc. Each is selected according to the raw material.

MICROS mill has the following advantages:

- Compactness and easy operation.
- Shaft seal section is constructed so that the structure can be sealed with an inert gas such as $N_2$ or Ar, etc. for easy control atmospheric conditions inside the equipment.
- Since the vessel is equipped with a jacket, heating or cooling temperature operation is possible.
- Any gas generated by the powder reaction during dry pulverization can be easily discharged outside of the system.
- Grinding media is selectable from stainless steel, ceramic or super hard material, etc.
- An invertor eases the control of the revolution setting.

Nara Machinery Co., LTD produces the mill with diameter of working vessel from 87 to 378 mm, with power of motor from 2,2 to 55 kW, and with rotation speed from 250 to 3000 $min^{-1}$.

MICROS mill can operate both in batch and continuous regime; vertical and horizontal types of working vessel arrangement are possible. Ring roll mills were successfully used for mechanochemical synthesis of $Ca_3(PO_4)_2 \cdot nH_2O$ from $Ca(H_2PO_4) \cdot H_2O$ and $Ca(OH)_2$ [12].

***References***


1. Bushuev L.P. On design and application of planetary mills. Izvestiya Vusov, Gornii Zhurnal 1960; 2: 17-20.
2. Bushuev L.P. Multiregime planetary mill. Izvestiya Vusov, Mashinostroenie 1964; 10: 17-22.
3. Patent RU №101874, 1955. Centrifugal planetary mill. Golosov S.I
4. Golosov S.I., Molchanov V.I. *Physico-chemical changes of minerals during fine grinding*. Novosibirsk: Institute of Geology and Geophysica, 1966.
5. Patent RU № 975068, 1982. Planetary mill. Avvakumov E.G.,Potkin A.R., Samarin O.I.
6. Patent RU № 158203, 1993. Planetary mill. Avvakumov E.G.,Potkin A.R., Bereznyak V.M.
7. Avvakumov E.G. *Mechanical Methods of Activation of Chemical* Processes. Novosibirsk: Nauka, 1986.
8. Gerasimov K.B., Gusev A.A., Kolpakov V.V., Ivanov E.Yu. Measuring of background temperature of mechanical alloying in planetary mill. Sibirskii Khim. Zhurnal 1991; 3: 140-45.
9. Avvakumov E.G. Mechanical activation of reactions between solid inorganic substances. Proc.V Symposium TATARAMAN, Bratislava, 1988; 2: 21-26.
10. Borunova A.B., Zhernovenkova Yu.V., Streletskii A.N., Portnoy V.K. Determination of energy intensity of mechanochemical reactors of different types. Book of Abstracts of INCOME-3, Prague, 4-8 September, 2000.
11. Patent RU № 2001680, 1993. Vibro-centrifugal mill. Denisov M.G., Denisov G.A., Novikov G.N.
12. Liao J., Ono K., Kanayama G., Isobe T., Senna M. Preparation of tricalcium phosphate by applying grinding technique and choosing hydroxycontaining initial materials. Chemistry for Sustainable Development 1998; 6: 233-36.

# Chapter 5

# MECHANICAL ACTIVATION OF HYDRATED OXIDES

## 5.1. Ordinary hydrated oxides

Many studies deal with the influence of mechanical activation on the properties, structure and composition of hydroxides.

**Magnesium hydroxide**. A detailed investigation of the influence of mechanical activation on the structure, thermal decomposition and dissolution of magnesium hydroxide was carried out in [1]. It was demonstrated that specific surface area of activated magnesium hydroxide decreases as a result of the formation of secondary particles (agglomerates); then it remains constant ("negative grinding"). However, the amorphization degree of the product increases, at first rapidly, then reaching some constant level at about 40 %. According to X-ray analysis, the size of crystallites, determined by Scherer's equation, at first decreases rapidly from 15–20 to 5–7 nm, then becomes constant. No mechanochemical decomposition of magnesium hydroxide was observed; it remained stable even under subsequent thermal treatment up to 150°C.

The data on thermal decomposition of activated magnesium hydroxide are well described by the first-order equation. The rate constant increases by a factor of 5–7 while the time of magnesium hydroxide activation in a vibration mill increases till 6 h. An increase of the decomposition rate is due to the fact that the nuclei formation of a new phase is simplified in the case of amorphous phase not only on the surface but also in the bulk.

The data on the dissolution of activated magnesium hydroxide are well described by the equation of shrinking sphere:

$$1 - (1 - \alpha)^{1/3} = K_d t \qquad (5.1)$$

In this case, on contrary, the constant of the equation decreases versus activation time, since the dissolution rate, determined by the outer surface of secondary particles, decreases in the course of activation.

**Aluminium hydroxides**. Aluminium trihydroxide is known to have three modifications: hydrargillite (gibbsite), bayerite and norstandite. Most widely used gibbsite is a natural mineral; it can be also synthesized by decomposition of aluminate solutions according to Bayer's technique. The density of gibbsite is 2.30–2.43 g/cm$^3$, its hardness is 2.50–3.75 (Mohs hardness scale).

Gibbsite has a layered structure; each layer is composed of two sublayers incorporating hydroxyl ions parallel to the (001) plane and a layer of aluminium ions being located among them. Aluminium ions are in the centres of octahedrons composed of $OH^-$ groups. Octahedrons are linked by common edges to give the rings $Al_6(OH)_{24}^{6-}$ with channels inside them. The ensemble of rings with common hydroxyl groups gives triple layers in hydrargillite lattice. The triple layers are linked to each other so that each $OH^-$ ion of one layer is opposite to hydroxyl ion of the next layer. Triple layers are held each other only by hydrogen bonds.

It was demonstrated [2-4] that during the mechanical activation, gibbsite crystals are comminuted, their size being decreased to 200–1000 nm, while hydroxide packets are shifted, that is accompanied by rearrangement of the structure.

Due to structural disordering caused by activation, the substance gets excess energy. A maximum amount of energy (up to 33 kJ/mol $Al(OH)_3$) is accumulated at a moment when the thickness of plates reaches its minimum (~ 2 nm). The product as prepared is X-ray amorphous and highly active. It is dissolved several times faster than non-activated one.

The total composition of aluminium hydroxide after mechanical activation remains close to the formula $Al(OH)_3$; however, according to NMR spectroscopic data, substantial changes in the coordination of $Al^{3+}$ cation occur: the cations in tetrahedral and pentahedral coordiantion appear. They are formed as a result of the rupture of Al-OH bonds. If the layered character of the initial structure is conserved, two $OH^-$ groups surrounded aluminium ion interact with each other to form $O^{2-}$ and $H_2O$. Polyhedrons $[Al(OH)_2O_3]$ are formed as an intermediate state. If the protons of the remaining OH groups react with OH groups of the neighbouring polyhedrons, the polyhedron $[AlO_5]$ is formed along with water molecule. The following interaction of protons of different OH brings to tetrahedron $[AlO_4]$ formation. These polyhedrons are responsible for the $^{27}Al$ NMR signals at 36 and 67 ppm, respectively [5-8].

Aluminium hydroxide is known to undergo a sequence of following transformations under heating:

$$Al(OH)_3 \rightarrow \gamma\text{-}AlOOH \rightarrow \gamma\text{-}Al_2O_3 \rightarrow \delta\text{-}Al_2O_3 \rightarrow \alpha\text{-}Al_2O_3.$$

The coordination of aluminium is changed during the transformation from octahedral to mixed coordination in $\gamma$ - $Al_2O_3$ (the latter has a spinel-like structure containing 16 octahedral aluminium and 8 tetrahedral positions) and in $\delta$-$Al_2O_3$ (containing some elements of spinel lattice), then again octahedral in $\alpha$-$Al_2O_3$ [8].

Mechanical activation brings to a substantial decrease of transition temperature and the decrease of a number of transitions:

$$Al(OH)_3 \rightarrow \gamma - AlOOH \rightarrow \gamma - Al_2O_3 + \alpha - Al_2O_3 \rightarrow \alpha - Al_2O_3.$$

Mechanical activation of boemite, $\gamma$-AlOOH, for 1.5 h can lead to the formation of $\alpha$ - $Al_2O_3$ without heating [8].

The changes in the coordination number of aluminium after activation and following heat treatment of gibbsite are presented in [7]. It was shown that penta-coordinated aluminium atoms arise under mechanical activation while under thermal treatment they are absent. This fact stresses the difference between mechanochemical and thermal processes. Besides, the same authors [9] consider that penta-coordinated aluminium atoms arise on the surface of particles and, similarly to gel-like state, are stabilized by protons or water molecules. Penta-coordinated aluminium atoms are much more numerous on the surface of gibbsite than on the surface of boemite and $\gamma$-$Al_2O_3$, their appearance resulting in the increased reactivity of the surface of mechanically activated gibbsite. When duration of mechanical activation increases, oxolation degree of gibbsite (the formation of –Al-O-Al- bonds) becomes higher, bringing to decrease of reactivity, in particular solubility in acids and bases.

Bayerite, though having the same total formula as gibbsite, however, differs by the character of hydroxyl layer packing, and its behaviour under mechanical activation differs from that of gibbsite [10].

Bayerite has a monoclinic lattice, $P2_{1/a}$, its parameters being a = 0.562, b = 0.867, c = 0.471 nm, $\beta$ = 90.26° [10]. It is based on AB packet composed of two tightly packed layers of hydroxyl ions with a layer of aluminium cations located between them. The arrangement of packets in bayerite can be expressed as ABAB [8]. The packets are perpendicular to the axis C and held between each other by hydrogen bonds. The average distance between two layers in a packet is equal to 0.207 nm while the distance between two neighbouring layers from different packets is 0.260 nm.

Mechanical activation of bayerite is accompanied by a shift of hydroxyl packets followed by the splitting of crystals into plates down to 1 nm thick. While shifting, packet alternation changes from ABAB... to ABBA... which is characteristic of gibbsite, and ABCABC..., characteristic of cubic packing. Activation is accompanied by dehydroxylation with the formation of $O^{2-}$ and molecular water which is located in the first coordination sphere of $Al^{3+}$ cations. A transition of bayerite into X-ray amorphous compound $Al_2O_3 \cdot 3H_2O$ is observed, the latter having octahedral cation coordination and disordered cubic packing of $OH^-$, $O^{2-}$, $H_2O$ [10].

Boemite, $\gamma$ - AlOOH, is aluminium monohydroxide, its structure being composed of the layers of cubic tightly packed oxygen ions with aluminium ions localized in octahedral sites between ordered layers. Under heating, it is dehydroxylated to form $\gamma$-$Al_2O_3$, and then $\alpha$-$Al_2O_3$. Mechanical activation destroys long-range order; activation for a long time leads to the formation of $\alpha$-$Al_2O_3$ without any heating.

The authors [9] investigated the influence of mechanical activation on the structure and thermal behaviour of boehmite by X-ray powder diffraction, thermal analysis and $^{27}Al$ NMR. They showed that grinding of boehmite for 60 min does not cause the boehmite to dehydrate, but brings to decrease in intensity of its X-ray reflections, and the formation of an amorphous phase containing tetrahedral aluminium and some other aluminium site with a resonance at 36 ppm (sometimes attributed to pentacoordinated aluminium). The authors have found a correlation between this resonance and adsorbed water content changed under mechanical activation. The ground material thermally transforms to corundum at 940°C via $\gamma$-$Al_2O_3$, by contrast with the unground sample, which forms corundum at 1195°C via $\gamma$- and $\delta$-$Al_2O_3$.

**Hydrated silica**. As shown in [11], mechanical activation of hydrated silica $SiO_2 \cdot nH_2O$ ($n = 0.4 \div 1.0$) brings to the removal of water up to 0.3–0.4 moles per 1 mole of $SiO_2$. The product is X-ray amorphous and highly dispersed. Its heating at 1100°C is accompanied by a decrease of specific surface area to 2–3 $m^2/g$. Water molecules and $OH^-$ groups present on the surface are removed under heating, either due to the condensation of the surface $OH^-$ groups, or due to the surface diffusion of a proton of Si-OH group. At 1070°C, the absorption band at 3680 $cm^{-1}$, attributed to $OH^-$ groups in the interior of solid particles, disappears completely, pointing to its practically complete dehydroxylation. However, some $OH^-$ groups still remain on the surface after heating at this temperature.

**Titanium and zirconium hydroxides.** It is difficult to obtain these hydroxides in crystalline form, therefore, the effect of mechanical activation on dehydration and crystallization of amorphous gels at room temperature was studied in [12].

The change of specific surface, water content in gels and crystallization temperatures are shown in Table 5.1.

As it follows from Table 5.1, under mechanical activation, a substantial dehydration of gels (up to 50 %) occurs; crystallization temperature changes: for $TiO_2$ gel, it decreases, while for $ZrO_2$, gel it increases. Besides, according to X-ray analysis and Raman spectroscopy, mechanically activated samples of $TiO_2$ gels are characterized as anatase, while $ZrO_2$ gels - as both tetragonal and monoclinic modifications.

The authors explain the mentioned differences by the difference in $TiO_2$ and $ZrO_2$ structures. Anatase and ruthyl structures possess the tightest cubic packing with the coordination number 6, while $ZrO_2$ structure is not tightly packed; its coordination number is 7 or 8. Therefore, the latter structure can incorporate more hydroxyl groups than $TiO_2$.

*Table 5.1.* The properties of initial and activated gels of titanium and zirconium dioxides [12].

| Oxide | Treatment | S, $m^2/g$ | Water content, mol $H_2O$/mol oxide | $T_{cryst}$, K | $T_{init}$ – $T_{act}$, K |
|---|---|---|---|---|---|
| $TiO_2$ | Gel | 325 | 1.1 | 698 | - |
| | Rotational mill, 3 h | 78 | 0.8 | 680 | 18 |
| | Rotation mill, 10 h | - | 0.7 | 688 | 10 |
| | Vibrational mill, 3 h | 16 | 0.6 | 654 | 44 |
| $ZrO_2$ | Gel | 177 | 1.7 | 692 | - |
| | Rotational mill, 3 h | 99 | 1.2 | 706 | -14 |
| | Vibrational mill, 0.5 h | - | 0.8 | 710 | -18 |

It is known that the tetragonal phase is formed at first in $ZrO_2$ gel under heating, since there are similar coordinations of $Zr^{4+}$ ion in gel and in the tetragonal phase (z = 8). Later on, a less coordinated asymmetric monoclinic modification is formed (z = 7). This causes strengthening of Zr – OH, Zr – O bonds in the monoclinic phase. In the course of mechanical activation, the rearrangement of bonds occurs in a similar way: gel → tetragonal → monoclinic. It is confirmed by the appearance of X-ray reflections from tetragonal and monoclinic modifications.

The strengthening of bonds in the monoclinic modification causes an increase of the crystallization temperature of activated gels. The activation of $TiO_2$ gel causes anatase formation and simplifies further transformation into rutile ($T_{crys}$ decreases).

## 5.2. Hydrated complex oxides

**Talc.** Thanks to its useful properties (thermal and chemical stability, lubricant ability, etc.), talc is one of the most important magnesium silicates. After dehydration it turns into enstatite which is widely used in ceramic industry.

Talc is described by the formula $Mg_3Si_4O_{10}(OH)_2$, or ($3MgO{\cdot}4SiO_2{\cdot}H_2O$); it possesses monoclinic syngony with cell parameters a=0.527, b=0.912, c=1.885 nm, β=100°00′, z=4. Its density is 2.7–2.8 g/cm$^3$ and a hardness 1 (Mohs hardness scale).

Mechanical activation of talc exhibits some specific features connected with the changes in the state of $H_2O$ and OH groups and caused by crystallographic degradation of talc [13].

Its structure is three-layered, i.e., it is formed from two hexagonal silica layers and one central brucite layer between them. There are four types of chemical bonds in talc structure: ionic Mg-O bonds, covalent, partially ionic Si–O and H–O bonds; hydrogen bonds between OH and O–Si, and van der Waals bonds between the layers. Van der Waals bonds are the weakest among all bonds. Therefore, early stages of grinding are accompanied by mechanochemical dehydration which can be represented by the equation:

$$2(OH)^- \rightarrow (1 - m)\,H_2O + (1 - m)\,O^{2-} + 2m(OH)^- \qquad (0<m<1)$$

Dehydration results in the rupture of Mg–O and $OH^-$–(O–Si) bonds. According to equation, a part of oxygen ions (1–m) is released as a result of mechanical activation. Covalent Si-O bonds are the most difficult to be destroyed. That is why $SiO_4$ tetrahedrons are very stable; only the bonds between the oxygen ions occupying the angles and providing binding between two tetrahedrons are broken, giving broken bonds. Energy is accumulated mainly in the outer Si – O layers; the central part amorphizes and undergoes mechanochemical dehydration mentioned above.

Under heating, activated talc is transformed into enstatite:

$$Mg_3Si_4O_{11-m}(OH)_{2m} \rightarrow 3MgSiO_3 + SiO_2 + mH_2O\,(0<m<1).$$

The transformation is accompanied by re-orientation of the bound $SiO_4$ tetrahedrons. The formation of free oxygen ions $O^{2-}$ during activation simplifies enstatite formation since free oxygen ion recombines with partially destroyed bonds of $SiO_4$ tetrahedrons to give chains characteristic of enstatite. Thus, the limiting stage of enstatite formation from talc is the formation of a new Mg–O–Si order with highly mobile $O^{2-}$ ions formed during mechanical activation.

**Kaolinite.** The composition of this mineral is close to the empiric formula $Al_2O_3{\cdot}2SiO_2{\cdot}2H_2O$ while structural chemical formula of aluminium silicate is $Al_2[Si_2O_5](OH)_4$. It has triclinic syngony, P1 space group, unit cell parameters are: a = 0.515, b = 0.895, c = 0.737 nm, $\alpha$ = 91°48′, $\beta$ = 104°30′, $\gamma$ = 90° (z = 2).

Kaolinite structure is composed of a layer incorporating two networks: tetrahedral network made of tetrahedrons $[SiO_4]$ and octahedral network made of octahedrons $[Al(O,OH)_6]$. The latter is similar to network in gibbsite structure, $Al(OH)_3$, with nearly equal lengths of corresponding a and b axes. The networks unite at an insignificant change of interatomic distances. Such structure provides a sharply polar character of the layer. When packed into a structure, the layers are bound to each other by very weak bonds, that is why their structure is always not completely ordered.

Mechanical activation of kaolinite was investigated in [14-19]. It was stated that at the initial stages of activation kaolinite undergoes a substantial increase of the number of active surface sites of acidic and basic character. Their concentration increases from 0.9 mmol/g in the initial sample to 1.3 mmol/g in the samples activated for 0.5 h; then it decreases with increasing activation time. These sites are of different nature, e.g., silanole groups Si–OH, exhibiting weakly acidic character (with $pK_a$ 6.5–9.5), or strongly acidic sites of Lewis type arose due to coordination-unsaturated metal cations (mainly aluminium) on the surface [17].

Kaolinite is transformed into X-ray amorphous state when activated in air. According to authors [14,15], amorphization involves the destruction of bonds between tetrahedral and octahedral layers inside the package, till the decomposition into amorphous aluminium and silicon oxides. Other researchers [16,17] consider that amorphized kaolinite conserves the initial ordering of the positions of silicon atoms while disordering of the structure is due to the rupture of Al – OH, Si – O – Al bonds and the formation of molecular water. Endothermic effect of the dehydration of activated kaolinite is shifted to lower temperatures while intensive exo-effect with a maximum at 980°C still conserves. When mechanically activated kaolinite annealed at 1000°C, only mullite ($3Al_2O_3 \cdot 2SiO_2$) and X-ray amorphous $SiO_2$ are observed. In this case, the phase with spinel structure which is formed under thermal treatment of non-activated kaolinite is not observed; thus, mechanical activation leads to the formation of other phases.

According to electron microscopic data, the initial kaolinite sample is composed of thin plates (6–30 nm) shaped as pseudo-hexagons with cross-section size of 0.1–2 μm. After activation, kaolinite is composed of the plates with rounded edges and cross-section size up to 0.6 μm, shifted and turned around other plates. According to IR spectroscopy, at first, Al–OH bonds (absorption bands 910 and 938 $cm^{-1}$) and Al-O-Si bonds (absorption bands 540 $cm^{-1}$) are broken [18].

Important data on the changes in the coordination of aluminium cations in kaolinite structure were obtained by NMR [17-19]. Non-activated sample exhibits chemical shift on $^{27}Al$ ($\delta$ = -2 ppm) corresponding to Al coordination number 6. After activation, the signals with chemical shifts $\delta$ = 53, 29, -2 ppm corresponding to 4, 5 and 6-coordinated states of aluminium ions are observed. The change of chemical shift of aluminium ions is due to dehydroxylation with the formation of molecular water, which remains within activated kaolinite bulk. It is either bound to aluminium atoms (Al–OH), supplementing their surroundings to octahedrons, or hydrogen-bound to particle surface. The heating of activated samples increases the amount of 5-coordinated aluminium. At increased temperature, the resulting meta-kaolinite rearranges into mullite and amorphous silica.

Thus, under mechanical activation, new kaolinite states are formed, which are not observed at thermal treatment.

**Silicate phases of granite.** Most of natural materials are multicomponent heterophase mixtures of minerals, for example, granites are composed of the silicates of different structures. The main types of crystal structures in silicate phases of granite powder are framework, layered and chain-like structures built of $SiO_4$ tetrahedrons.

The authors [20] studied the influence of mechanical activation on the changes in the composition of silicate phases of granite powder which is a large-scale industrial waste material. The powder incorporates the following phases: quartz - $SiO_2$, augite - Ca(Fe, Mg)$SiO_6$, laumontite - $CaAl_2Si_4O_{12}\cdot 4H_2O$, hornblende - $(Na, K, Ca)_x(Al, Mg)_7(Si, Al)_8O_{20}(OH)_{10}\cdot xH_2O$, xonotlite - $Ca_6Si_6O_{17}(OH)_2$, enstatite - (Mg, Fe)$SiO_3$ and small amounts of other phases. Its average chemical composition is (mass. %): $SiO_2$ – 53.9, $Al_2O_3$ – 15.5, CaO – 8.5, $Fe_2O_3$ – 7.2, MgO – 4.6, $Na_2O$ – 2.4, $K_2O$ – 1.6, $TiO_2$ – 0.5, $MnO_2$ – 0.2, $B_2O_3$ – 0.1.

Mechanical activation brings to a decrease of the intensity of X-ray reflections of all the phases except quartz. Permanent presence of $SiO_2$ phase can be the sequence of its stability to deformations as well as the result of its formation from mechanically decomposed silicates. The initial phase, $SiO_2$, serves as a nuclei for its growth from amorphous phases during mechanical activation of silicates.

The intensities of X-ray reflections of laumontite, $CaAl_2Si_4O_{12}\cdot 4H_2O$; hornblende, $(Na,K,Ca)_x(Al,Mg)_7(Si,Al)_8O_{20}(OH)_{10}\cdot xH_2O$; xonotlite, $Ca_6Si_6O_{17}(OH)_2$, decrease after a short-time mechanical activation and practically disappear after average mechanical load. The intensities of reflections of phases comprising plagioclase (albite, $NaAlSi_3O_8$, and anortite, $CaAl_2Si_3O_8$), audite, Ca(Fe,Mg)$Si_2O_6$, and enstatite, (Mg,Fe)$SiO_3$, also decrease while the width of reflections is broadened versus time of mechanical treatment. In the case of the strong regime of mechanical activation, the initial material becomes X-ray amorphous; the traces of plagioclase, enstatite and quartz phases are observed. Thus, mechanical activation causes the destruction of chain-like hydrosilicates, opposite to framework-structure compounds which are more stable.

Silicates contain structural and adsorbed water. The major part of water is evolved from the initial material (~ 67 %) under heating at 250–740°C. It is known that the dehydration of silicates is accompanied by the destruction of silicate structure. Really, laumontite, $CaAl_2Si_4O_{12}\cdot 4H_2O$, hornblende, $(Na,K,Ca)_x(Al,Mg)_7(Si,Al)_8O_{20}(OH)_{10}\cdot xH_2O$, xonotlite, $Ca_6Si_6O_{17}(OH)_2$, disappear after the heating of the activated powder at 1000°C. Anorthite, albite, enstatite, hematite, $SiO_2$ phases and those similar to $Ca_{1,82}Al_{3,64}Si_{0,36}O_8$ are formed under heating.

Mechanically activated samples release increased amounts of water under heating at 200°C. This can be explained by the fact that mechanical activation brings as well as to disordering of the initial silicate structures as to their decomposition with the formation of new compounds accompanying dehydration. Absorption bands at 3800–3600 $cm^{-1}$ which correspond to OH group are practically absent in IR spectra.

The products obtained by thermal treatment of mechanically activated granite are substantially different from those for non-activated granite.

The joint activation of granite powder with calcium oxide CaO was carried out in order to bind them together into the minerals of cement clinker: di- and tri-calcium silicates. After annealing of mixtures, the formation of aluminosilicates, $2CaO{\cdot}Al_2O_3{\cdot}SiO_2$ and $CaO{\cdot}Al_2O_3{\cdot}2SiO_2$, the crystallization of $Ca(Mg,Al,Fe)Si_2O_6$ and the formation of the compounds of $MgO{\cdot}Al_2O_3{\cdot}3SiO_2$ type was observed.

Thus, mechanical activation of a mixture of silicate phases of granite leads to their structural decomposition. Layered phases are destroyed more rapidly than framework ones. Besides, adsorbed water is removed at lower temperature (200–300°C). Thermal treatment of mechanically activated mixture with CaO leads to the formation of minerals of cement clinker.

## *References*


1. Inone S. I., Senna M. Differences between the dissolution and the decomposition kinetics of mechanically activated magnesium hydroxide. Reactivity of Solids 1988; .5: 155–66.
2. Paramzin S.M., Pankratyev Yu.D., Pauishtis E.A. et al. Studies of the products of mechanical activation of hydrargillite. I. The state of water in activated samples. Izv. SO AN SSSR, ser. khim. nauk, 1984; 4: 33–38 .
3. Paramzin S.M., Krivoruchko O.P., Zolotovsky B.P. et al. II. Morphology, structure and chemical activity of mechanochemically treated samples. Izv. SO AN SSSR, ser. khim. nauk, 1984; 6: 39–42.
4. Paramzin S.M., Pankratyev Yu.D., Turkov V.M. et al. III. Accumulation of energy during mechanochemical activation of Al(III) trihydroxides. Izv. SO AN SSSR, ser. khim. nauk, 1988; 2: 47–50.
5. Krivoruchko O.P., Mastikhin V.M., Zolotovsky B.P. et al. A new coordinaton state of Al(III) ions in aluminium hydroxides. Kinetika i Kataliz 1985; 26: 763.
6. Krivoruchko O.P., Paramzin S.M., Plyasova L.M. A new Al(III) hydroxide with a composition $Al_2O_3$ $3H_2O$ and cubic packing of oxygen. Kinetika i Kataliz 1987; 28: 765.
7. Mackenzie K.I.D., Temunjin J., Okada K. Thermal decomposition of mechanically activated gibbsite. Thermochimica Acta 1999; .327: 103–08.
8. Zolotovsky B.P., Klevtzov D.P., Paramzin S.M., Buyanov R.A. The synthesis of complex oxides using mechanical activation. In *Mechanochemical Synthesis in InorganicCchemistry*, ed. Avvakumov E.G, Novosibirsk: Nauka, 1991.
9. Paramzin S.M., Plyasova L.M., Krivoruchko O.P. et al. Investigation of structural transformations of bayerite during mechanochemical activation. Izv. SO AN SSSR, ser. khim. nauk 1988; 6: 1209–13.
10. Mackenzie K.I.D., Temunjin J.,.Smitnetd M.E, Effect of mechanochemical activation on the thermal reactions of boehmite (γ-AlOOH) and γ-$Al_2O_3$. Thermochimica Acta 2000; 354: 87–94.
11. Kosova N.V., Avvakumov E.G. Mechanochemical synthesis of calcium silicates based on hydrated oxides. Sib. Khim. Zhurn. 1992; 2: 135–43.
12. Isobe T, Senna M. Change in dehydration and crystallization of inorganic gels due to mechanical activation. Sib. Khim. Zhurn. 1991; .5: 25-31.
13. Liao I,.Senna M. Thermal behavior of mechanically amorphized talc. Thermochimica Acta 1992; 197: 295–306.
14. Takahachi H. Wet grinding on kaolin minerals. Bull. Chem. Soc. Japan 1959; 32: 235-54.
15. Schrader R., Kutzer H, Hoffman B. Uber der mechanische activirung von kaolinit. Tonind. Zeit. 1970; 94: 410–16.
16. Lapteva E.S., Yusupov T.S., Berger A.S., *Physicochemical Changes of Layered Silicates during Mechanical Activation*. Novosibirsk: Nauka, 1981.
17. Kornilovich, Boris, *Structure and Pproperties of Mechanically Activated Silicates and Carbonates*. Kiev: Naukova dumka, 1994.
18. Klevtsov D.P., Mastikhin V.M., Krivoruchko O.P. Investigation of the mechanism of solid-phase transformations during thermal treatment of aluminosilicate systems. II. Investigation of the effect of mechanical activation on kaolinite state. Izv. SO AN SSSR, ser. khim. nauk, 1986; 3: 62–70.
19. Klevtzov D.P., Krivoruchko O.P., Mastikhin V.M. et al. Influence of mechanochemical activation and thermal treatment of kaolinite and cation distribution of Al(III) and formation Na–A zeolite. React. Kinet. Catal. Letter 1988; 36: 319–24.
20. Lomovsky O.I., Golubkova G.V. Mechanochemical transformations in mixture of the silicate phases of granite. J. Mater. Synthesis Processing 2001 (to be published).

# Chapter 6

# MECHANOCHEMICAL SYNTHESIS OF DOUBLE OXIDES

The syntheses of double, triple and more complicated compounds can be carried out on the basis of hydroxides and hydrated oxides. In order to achieve the most efficient interaction, one should select initial components taking into account their acid-base properties, solubility and possible hydrolysis in the presence of small amounts of water. The compounds or groups of compounds should be selected to provide a maximum difference in acid-base properties. A quantitative characterization of these properties was described in the previous chapters. By present, a substantial experience in soft mechanochemical synthesis of inorganic compounds has been accumulated. The description of experimental results presented below is divided into two parts. One of them deals with the synthesis of double compounds and an another with the synthesis of more complicated compounds.

## 6.1. Aluminates

The methods of mechanochemical synthesis of various aluminates, including double compounds of alkaline, alkaline earth and transition metals, and aluminium, have been developed. The data on their synthesis, based on aluminium hydroxide, are presented according to this sequence.

**Lithium aluminates.** The compounds $LiOH\cdot 2Al(OH)_3\cdot mH_2O$ and $LiCl\cdot 2Al(OH)_3\cdot mH_2O$ (here m = 0.5; 1.0; 2.0) are easily synthesized under very low mechanical loading (blade mixer) in stoichiometric $Al(OH)_3$+$LiOH\cdot H_2O$ and $Al(OH)_3$+$LiCl\cdot H_2O$ mixtures [1,2]. It was stated that the dispersion of the initial aluminium hydroxide strongly influenced on the kinetics of mechanochemical interaction. The interaction rate increases linearly with the specific surface of the initial hydroxide. Fig. 6.1 shows the data on reactivity of initial hydroxide with different specific surface area (6 and 2 $m^2/g$, respectively).

The heating of the products as prepared up to 200°C is accompanied by the loss of interlayer water; above 200°C, the decomposition of $LiOH\cdot 2Al(OH)_3\cdot mH_2O$ takes place yielding anhydrous lithium aluminate

$Li_2O \cdot Al_2O_3$; $LiCl \cdot 2Al(OH)_3 \cdot mH_2O$ decomposes to give lithium chloride and aluminium oxide.

Double compounds of lithium and aluminium are used to extract lithium from aqueous salt-containing solutions [3]. They decompose in water releasing lithium compounds.

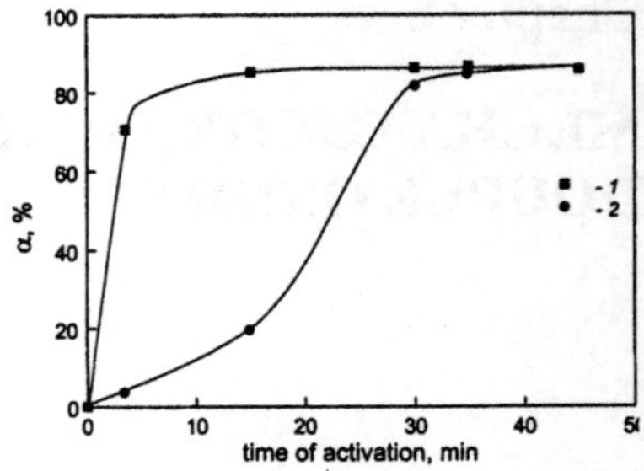


*Figure 6.1.* Time dependence of the interaction degree in the mixture $Al(OH)_3$+$LiCl \cdot H_2O$. 1- fine particles; 2 – coarse particles.

**Magnesium aluminates.** The spinel $MgAl_2O_4$ is commonly used as a refractory material in heavy industry because of its excellent mechanical and thermal properties at elevated temperatures. Its solid state synthesis from magnesia and alumina occurs via interdiffusion of cations through the product layer between the oxide particles, that needs high processing temperatures (>1400÷1500°C). It would therefore be desirable to obtain the magnesium aluminate precursor by mechanochemical processing; this could then be converted at about 1000-1050°C to spinel with a high surface area and low-sized particles to achieve good sintering behaviour.

The author [4] observed the formation of a new phase (probably, double hydroxide of aluminium and magnesium) under mechanical activation of a mixture of magnesium and aluminium hydroxides in a planetary mill. The mentioned product decomposes at 800°C to form spinel $MgAl_2O_4$.

Soft mechanochemical synthesis and sintering of magnesium aluminate spinel were investigated also in [5]. X-Ray amorphous precursor phases for the synthesis of $MgAl_2O_4$ have been prepared by grinding of $Al(OH)_3$+$Mg(OH)_2$ mixture or $4MgCO_3 \cdot Mg(OH)_2 \cdot 4H_2O$. The mechanochemical treatment does not remove any water or carbonate, but, accordint to $^{27}Al$ MAS NMR, converts a part of octahedral Al sites in gibbsite into tetrahedral sites and some other sites with a resonance at about 38 ppm. The brucite-derived precursor forms spinel on heating at ≤ 850°C, by contrast to the unground mixtures which show insignificant spinel formation even at 1250°C. The brucite-derived spinel is sintered to 72% theoretical density and shows a morphology consisting of widely disparate grain sizes. At about 850°C, the hydromagnesite-derived precursor transforms into a mixture of spinel and hydrotalcite ($Mg_6Al_2(OH)_{16}CO_3 \cdot 4H_2O$), the latter decomposing to spinel and MgO at 1050°C. The ceramics derived from hydromagnesite-containing precursor shows superior pressureless sintering properties at 1400-1600°C, producing a body of 97% theoretical bulk density at 1600°C.

The anion forms of double hydroxides of magnesium and aluminium have been obtained by mechanical activation of crystal magnesium with aluminium salts in a planetary mill AGO-2 for 15 min according to the reaction:

$$3Mg(OH)_2 + Al(A_n)_3{\cdot}6H_2O = [Mg_2Al(OH)_6]A_n{\cdot}n\ H_2O + Mg(A_n)_2.$$

Here $A_n$ is $Cl^-$, $NO_3^-$, $SO_4^{2-}$, etc. [6].

Thus synthesized forms of double magnesium-aluminium hydroxide appeared to be good anion exchangers; they can be also used for the preparation of spinel $MgAl_2O_4$ by heating at temperature not above 800°C by adding aluminium hydroxide to achieve spinel stechiometry.

By means of mechanical activation of $3MgCO_3$+$Al(OH)_3$ mixture in water solution (5% $Na_2CO_3$) for 24 h in vibratory mill, hydrotalcite, $Mg_6Al_2(OH)_{16}{\cdot}CO_3{\cdot}4H_2O$, was obtained [7]. Thermal decomposition of this product results in the preparation of fine-grained and high-strength spinel-magnesial ceramics.

Mechanochemical synthesis of anionic forms of hydroxides possesses some advantages over the traditional procedure of their preparation from solution. The compounds obtained by mechanochemical synthesis are sufficiently well crystallized and free of impurities. This synthesis method is ecologically safe.

**Calcium aluminates.** A wide range of calcium hydroaluminates exists in the system $CaO$-$Al_2O_3$-$H_2O$ with molar ratio $CaO{:}Al_2O_3$ changing from 1 to 6, while molar ratio $H_2O{:}Al_2O_3$ changes from 6 to 33 [8]. The compounds most stable under hydrothermal conditions are $3CaO{\cdot}Al_2O_3{\cdot}6H_2O$ and $4CaO{\cdot}3Al_2O_3{\cdot}3H_2O$. The former compound, represented by structural formula $Ca_3Al_2(OH)_{12}$, exists till 270-280°C, the latter is stable till 680-690°C. Dehydration of hydroaluminates can lead to anhydrous aluminates of various compositions.

The application of soft mechanochemical synthesis to the synthesis of calcium hydroaluminates is described in a series of works [9-14]. The author [4] investigated the effect of the nature of initial aluminium compounds, such as $\alpha$-$Al_2O_3$, $\gamma$-$Al_2O_3$, AlOOH (boemite) and $Al(OH)_3$ (gibbsite), on mechanochemical synthesis of calcium aluminate in the mixtures with calcium oxide. The highest rate was achieved with gibbsite. The rate decreases in the row $Al(OH)_3 \rightarrow AlOOH \rightarrow \gamma\text{-}Al_2O_3 \rightarrow \alpha\text{-}Al_2O_3$, that agrees with the decrease of acidic properties of these compounds.

It was demonstrated in [9] that mechanical activation of a mixture of $Ca(OH)_2$ with $Al_2O_3$ (at molar ratio 1:4) in a vibratory mill results in partial interaction with the formation of calcium hydroaluminate $4CaO{\cdot}3Al_2O_3{\cdot}H_2O$. The authors [9] showed that mechanochemical synthesis of $3CaO{\cdot}Al_2O_3{\cdot}H_2O$ occurs starting from $Ca(OH)_2$+$Al(OH)_3$ and $Ca(OH)_2$+ kaolinite mixtures during their activation at room temperature.

As described in [11], the same compound was synthesized by activation of the initial $3Ca(OH)_2$+$2Al(OH)_3$ mixture in a heated ball mill. It was demonstrated that the activation for 100 h at 100°C is sufficient for the direct synthesis of $3CaO{\cdot}Al_2O_3{\cdot}6H_2O$. At higher temperatures this compound is decomposed. It is also noted that the synthesis does not occur at given temperature without mechanical activation of the initial mixture of components.

The authors [12] used $Ca(OH)_2$+$4Al(OH)_3$ mixture for mechanochemical synthesis of $CaO{\cdot}2Al_2O_3{\cdot}7H_2O$. When heated to 1000°C, this compound decomposed to $CaO{\cdot}2Al_2O_3$.

The activation of mixtures of calcium and aluminum hydroxides taken at different ratios was studied in [13]. It was stated that $3CaO{\cdot}Al_2O_3{\cdot}6H_2O$ was the main product of mechanochemical synthesis within a wide range of component ratios (from 2:1 to 1:2). When a planetary mill AGO-2 is used, the synthesis completes within 10 min (Fig. 6.2). An increase of activation time brings to decrease of the intensity of reflections, which is an evidence of product amorphization or the beginning of hydroaluminate decomposition.

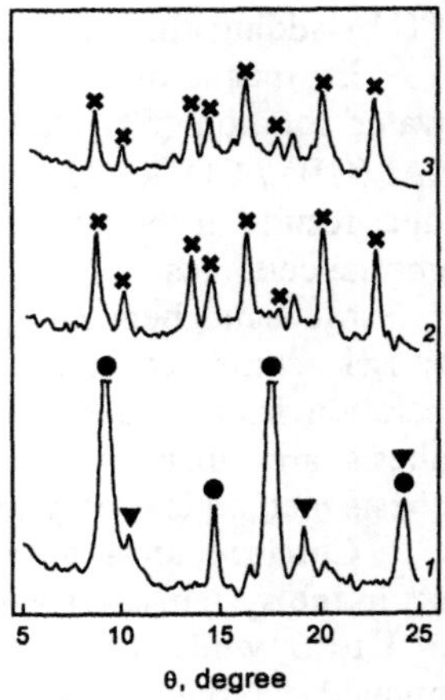


*Figure 6.2.* X-ray patterns of $Ca(OH)_2$+ $2Al(OH)_3$ mixture activated for: 5 (1), 10 (2), 15 (3) min.
● - $Ca(OH)_2$, ▼ - $Al(OH)_3$,
✖ - $3CaO{\cdot}Al_2O_3{\cdot}6H_2O$.

Alternative approaches exist to explaine the mechanism of chemical interaction. It is accepted that water is formed as a result of mechanochemical interaction, solid reagents are dissolved in water, and the reaction proceeds via the dissolved state (hydrothermal-like process). On the other hand, it is assumed that the initial stages of the process involve the interaction between acid and base centers present at the surfaces in contact, and the next stages are connected with the process of calcium cations insertion into aluminum hydroxide lattice.

The above-mentioned data [10] and those reported in [14] reveal the specific features of the process. In the latter paper it was demonstrated that $Al(OH)_3$ +CaO and $Al(OH)_3$+$Ca(OH)_2$ mixtures activated for 45 min in vibratory mill contained 5 and 7.5 mass.% of free water, respectively, due to dehydration of $Al(OH)_3$ (as the most easily decomposed component). It follows from [10] that mechanochemical synthesis in the same mixture does not proceed at temperatures above 100°C. The authors attribute the braking to the formation of less active product - boemite, AlOOH, but it is quite possible that the braking occurs due to the evaporation of water from the contact zones.

On the other hand, the structure of $Ca_3Al_2(OH)_{12}$, formed via interaction of $Al(OH)_3$ with $Ca(OH)_2$, consists of $OH^-$ ions packed with metal ions, which occupy the suitable size sites. Calcium cation is surrounded by eight $OH^-$ ions, while aluminium cation is surrounded by six $OH^-$ ions, corresponding distances being 0.250 and 0.192 nm, respectively. This means that the formation of structure can proceed in a simpler manner than through the dissolution followed by crystallization.

The authors [14] have received some aguments confirming the second mechanism. The concentrations of basic centres on CaO and $Ca(OH)_2$ after their activation either individually or in mixtures with $Al(OH)_3$ were determined by means of pH-titration. The number of centres per unit surface area increases in the first case. Vice versa, activation in mixtures leads to the decrease of the number of centres (Fig. 6.3). The decrease is most substantial in mixture with CaO. The residual concentration of centres is lower with CaO (~0.6 g-ion/m$^2$), in the case of $Ca(OH)_2$ it is 1.3 g-ion/m$^2$ . This means that basic centres are consumed during the activation to react with acidic centres of $Al(OH)_3$ surface, and the interaction efficiency correlates with basic properties: CaO > $Ca(OH)_2$.

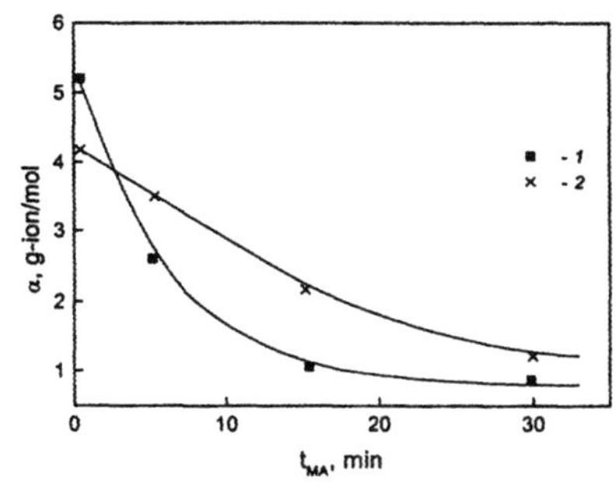


*Figure 6.3.* The changes of basic centre concentration versus time of activation of the mixtures:
1 – $Al(OH)_3$+CaO, 2 - $Al(OH)_3$+$Ca(OH)_2$.

Besides, the authors discovered an intercalation effect during activation. Intercalation involves the insertion of calcium ions into aluminium hydroxide lattice which possesses a layered structure. Intercalation takes place between hydroxide layers over the planes with Miller index {002}. A maximum increase of the distance is observed in $Al(OH)_3$+CaO mixture (from 0.4850 to 0.4875 nm); while in $Al(OH)_3$+ $Ca(OH)_2$ mixture, it changes not so significantly (to 0.4867 nm).

Anhydrous calcium aluminates obtained from calcium hydroaluminates by thermal treatment are used as the components of calcium-aluminate cement [15], as catalyst supports [16] and as sorbents for gas purification from fluorine [17].

**Barium hexa-aluminate.** $BaAl_{12}O_{19}$, being a compound of high practical importance, is usually synthesized via thermal decomposition of alkoxides.

Shortcomings of the method are the formation of large amounts of organic wastes and high temperatures of synthesis. The application of mechanical activation of solids allows to avoid liquid wastes during the synthesis and to decrease the temperature.

It was shown in [18] that practically monophase fine barium hexaaluminate can be obtained by mechanical activation of a mixture of barium oxide with $\gamma$-$Al_2O_3$, which exhibits acid properties to a larger extent than $\alpha$-$Al_2O_3$, and by consequent thermal treatments at increased temperature. The product then is grinded in the presence of water. The synthesis was shown to proceed almost completely after activation for 5 min in the AGO-2 planetary mill and thermal treatment at 1300°C for 1 h. Mechanical activation of the mixture of aluminium hydroxide with barium oxide, followed by thermal treatment at 900°C, results in the formation of the final product and $\alpha$-$Al_2O_3$ as an admixture which remains even at 1300°C. Mechanochemical synthesis helped also to synthesize barium hexaaluminate in which a part of aluminium cations is replaced with manganese, iron, cobalt cations. Such compounds are used as active ceramics in catalysis [17].

**Zinc aluminate.** Zinc hexaaluminate with atomic ratio Zn(II)/Al(III) = 3:1 and cell parameter c=0.75 nm (pyroaurite structural type) was obtained by mixing of ZnO crystal powder of hexagonal structure (with particle size ~0.2 micrometers) with $Al(OH)_3$ , preliminary activated in a planetary mill for 70 min, in distilled water [19]. The final product was formed during the ageing of the mixture for 24 h at 20°C. Non-activated $Al(OH)_3$ does not interact with ZnO (even if the latter is activated). The authors [19] consider the formation of zinc hexaaluminate as a specific solid phase reaction occured in disperse state via highly efficient mechanism involving the absorption of large-grain crystals of divalent metal hydroxides or oxides by the activated aluminium hydroxide.

**Lanthanum aluminate.** Mechanochemical synthesis of lanthanum aluminate was carried out by the joint grinding of lanthanum oxide $La_2O_3$ with aluminium hydroxide or oxide in a planetary mill [20]. After activation for 120 min, a monophase product $LaAlO_3$ with a large specific surface area was obtained. However, the formation of $LaAlO_3$ was not observed when $\alpha$-$Al_2O_3$ was used as an initial reagent.

## 6.2. Silicates

**Magnesium silicates.** Mechanochemical reaction between $Mg(OH)_2$ and $SiO_2$ was investigated in [21]. Activation was carried out in a laboratory vibratory mill. According to X-ray phase analysis data, for individually activated $Mg(OH)_2$ no changes in structure and composition are observed, except for the decrease of peak intensities due to amorphization. For activated mixtures of $Mg(OH)_2$ and $SiO_2$, amorphization proceeds more rapidly, practically within one hour. According to IR spectroscopy, an intensive absorption band at 3650 $cm^{-1}$ in the spectrum of activated mixture, assigned to the vibrations of isolated OH groups, disappears as early as after 2 h, while it is conserved in $Mg(OH)_2$, individually activated for 48 h. Similarly, the band at 790 $cm^{-1}$ in the spectrum of activated mixture, attributed to symmetrical vibrations of Si-O-Si bond, also disappears. Activation has no effect on bands at 3400 and 1630 $cm^{-1}$ referred to silanol group and adsorbed water. These results indicate that the mixture of $Mg(OH)_2$ and $SiO_2$ is physical only at early stages of grinding, then it is rapidly turned into a chemically bound complex.

These conclusions are in agreement with DTA and TG data. Endo-effect at 365°C, which is due to dehydration $Mg(OH)_2$, is observed for activated sample, too. Vice versa, it is absent for the activated mixture of $Mg(OH)_2$ with $SiO_2$, starting from 1 h. This means that dehydration occurs as a result of chemical interaction. The enthalpy of dissolution of this mixture passes through a maximum versus activation time, as in the case of activated talc, $Mg_3Si_7O_{10}(OH)_2$.

High-temperature treatment within the range 550-800°C brings to crystallization of products. In non-activated mixtures, heated at 800°, the crystallization of MgO and insignificant amounts of forsterite $Mg_2SiO_4$ are observed. In samples activated for a short time, a mixture of forsterite and enstatite is formed, while after activation for 3 h, only enstatite $MgSiO_3$ is formed.

The authors assume that, since mechanically activated silica has a high affinity to the hydroxyl groups of $Mg(OH)_2$, their interaction starts with the transfer of protons from $Mg^{2+}$ to $Si^{4+}$ ion. This process results in the formation of water molecules (from two OH groups) and $O^{2-}$ ions with a high mobility. These $O^{2-}$ ions participate in rearrangement of the $SiO_4$ bonds of the tetrahedron, which leads to the formation of a new chain structure of forsterite. The main statements of the model are as follows:

(1) the dehydration and formation of the precursor $mMgO{\cdot}nSiO_2$ take place at the same time at the contact points between unlike particles;
(2) the dehydration and formation of the precursor of magnesium silicates occur simultaneously in different local spots in a grinding vessel when the whole system is observed.

The rate of mechanochemical interaction is strongly affected by the particle size of initial components. Fine particles get in close contacts with each other; secondary aggregated particles arise, their size being equal to 2÷10 μm; chemical interaction proceeds with higher efficiency inside these particles.

Mechanochemical synthesis of $Mg_2SiO_4$, starting from different mixtures of anhydrous and hydrated magnesium and silicon oxides, was investigated in [22]. The goal of this study was to determine the effect of the amount of constitutional water in the hydrated silica on mechanochemical interaction in the systems 1) $Mg(OH)_2$ – silica gel, 2) $Mg(OH)_2$ – silicic acid, 3) MgO – silicic acid, and to find similarity with the products of hydrothermal synthesis of magnesium hydrosilicate.

The main sizes of initial particles of $Mg(OH)_2$, silicic acid and silica gel are 3, 12.4 and 20 μm, respectively, while specific surface area is 23.5, 142 and 580 $m^2/g$. Hydrothermal treatment of these mixtures was carried out at 80°C for 24 h. Hydrothermal reactions in $MgO\text{-}SiO_2\text{-}H_2O$ system were studied earlier by Yang [23], who stated the formation of talc, $3MgO{\cdot}4SiO_2{\cdot}H_2O$, and chrisotil, $3MgO{\cdot}2SiO_2{\cdot}2H_2O$, in this system. In X-ray diffraction patterns, basic reflections 0.94 nm and 0.737 nm correspond to these phases.

Mechanical activation of stoichiometric mixtures was carried out in a laboratory planetary mill at a rotation frequency of 400 rpm. According to X-ray data, after activation for 1200 min, no interaction was observed between the components in the first mixture, while in the second mixture, the interaction took place but it didn't complete; in the third case, the reaction ran to its completion.

According to DTA and TG data, endo-effect at 365°C, caused by $Mg(OH)_2$ decomposition, remains constant in the first case, becomes very small in the second case and is practically absent in the third case. This means that the interaction of initial components occurs at the stage of mechanical activation. DTA curves also exhibit endo-effect at 865°C, being maximal for the second mixture. It is due to the formation of forsterite, $Mg_2SiO_4$, in the first mixture and due to formation of a mixture of forsterite with enstatite in the second and the third mixtures.

According to IR spectroscopy, in the two latter mixtures, an absorption band at 950 $cm^{-1}$ corresponding to Si-O bond elongation in Si-OH group appears. According to $^{29}Si$ NMR data, these mixtures exhibit identical spectra with the parameters $Q^2$ = - 86 ppm and $Q^3$ = - 93 ppm, similarly to hydrothermally synthesized magnesium silicate. This spectrum is typical for hydrated magnesium silicate $3MgO{\cdot}2SiO_2{\cdot}2H_2O$. According to X-ray data, the final mechanochemical product has a main reflection at 0.84 nm, which can be attributed to talc or chrysotile-like products. This shows that there is a similarity between the products of mechanochemical and hydrothermal syntheses in this system. However, mechanochemical interaction does not take place in the second mixture due to large excess of water.

Interesting data were obtained by the same authors for the ageing of the mentioned mechanically activated mixtures [24]. According to these data, the interaction is not observed in the first mixture either after activation (30 min) or after ageing (6 months). The second mixture exhibited no interaction after activation for 30 min, but after ageing (6 months) the whole $Mg(OH)_2$ was bound to form layered silicate. According to $^{29}Si$ NMR data, peaks at -84,8 ppm and -92,5 ppm are observed; DTA curves exhibit endo-peak at 365°C.

Thus, it can be assumed that in the presence of small amounts of water, the interaction between magnesium hydroxide and silica, finely ground and thoroughly mixed, proceeds via the dissolution of silica in the presence of magnesium hydroxide. This followed by the interaction of silica with magnesium hydroxide resulting in the formation of insoluble magnesium silicate.

**Calcium silicates.** In [25] it was demonstrated that the reactions with the participation of hydrated silica are more profitable from thermodynamic viewpoint:

| | $\Delta G^\circ_{298}$, kJ/mol |
|---|---|
| $CaO + SiO_2 = CaSiO_3$ | -89,4 |
| $CaO + H_2SiO_3 = CaSiO_3 + H_2O$ | -173,5 |
| $Ca(OH)_2 + SiO_2 = CaSiO_3 + H_2O$ | -33,0 |
| $Ca(OH)_2 + H_2SiO_3 = CaSiO_3 + 2H_2O$ | -117,2 |

These mixtures, taken at a molar ratio of 1:1, were activated in a planetary mill; treatment time was varied from 1 min to 30 min. Chemical analysis showed that calcium was bound most efficiently in the second and fourth mixtures (Fig. 6.4). According to IR spectroscopy data (Fig. 6.5),

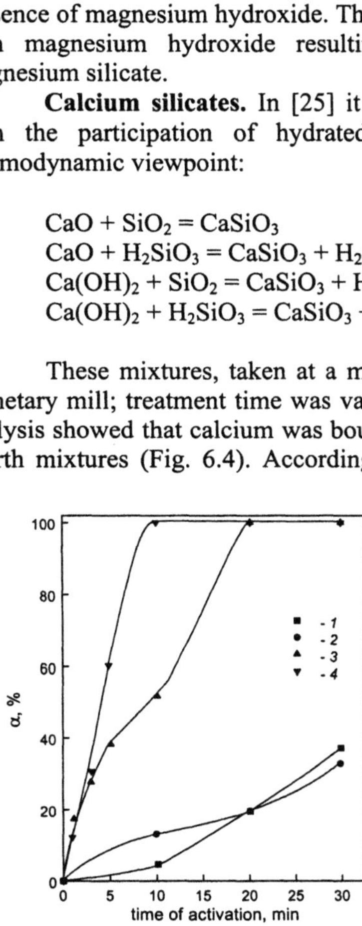


*Figure 6.4.* The kinetics of mechanochemical interaction in the mixtures: 1- $CaO+SiO_2$, 2 – $Ca(OH)_2+SiO_2$, 3 - $CaO+SiO_2{\cdot}0.6H_2O$, 4 – $Ca(OH)_2+SiO_2{\cdot}0.6H_2O$.

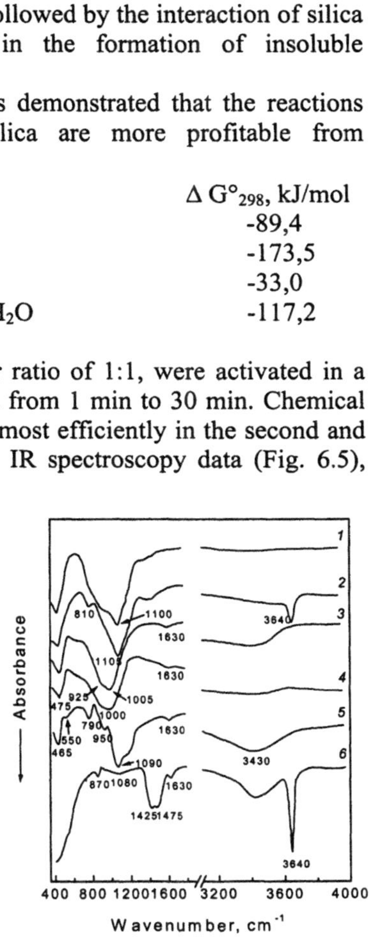


*Figure 6.5.* IR spectra of activated mixtures and initial compounds: 1- $CaO+SiO_2$, 2 – $Ca(OH)_2+SiO_2$, 3 - $CaO+SiO_2{\cdot}0.6H_2O$, 4 – $Ca(OH)_2+SiO_2{\cdot}0.6H_2O$, 5 - $SiO_2{\cdot}0.6H_2O$, 6 - $Ca(OH)_2$.

mechanical activation of mixtures under study brings to different changes of the bands corresponding to stretching Si-O bonds. In the spectra of $CaO+SiO_2$ and $Ca(OH)_2+SiO_2$ mixtures, together with the band at 1090 $cm^{-1}$, a shoulder at 925 - 940 $cm^{-1}$ appears. On contrary, in the spectra of $CaO+SiO_2{\cdot}0.63H_2O$ and $Ca(OH)_2+SiO_2{\cdot}0.63H_2O$, a band at 1090 $cm^{-1}$ is shifted to 1005 $cm^{-1}$, pointing to the significant $SiO_4$ rearrangement and the formation of new silicate phases. The spectrum of the activated mixture $Ca(OH)_2+SiO_2$ is characterised by the decrease of intensity of bands corresponding to the vibrations of isolated OH groups. In IR spectra of activated mixtures with $SiO_2{\cdot}0.63H_2O$, a broad asymmetric band appears in the range of 3600-2400 $cm^{-1}$, which is an evidence of a large amount of water molecules with strong hydrogen bonds in the structure of the products. This is typical for hydrated silicates.

On the basis of the data obtained, the authors conclude that mechanical activation of the mentioned mixtures leads either to partial interaction (in the case of anhydrous oxides) or to profound interaction (in the case of hydrated oxides). In the second case, the reaction completes with the formation of X-ray amorphous calcium hydrosilicates belonging to tobermorite group and being crystallized under heating in the form of vollastonite $\beta$-$CaSiO_3$.

Calcium hydrosilicates compose a wide range of different compounds [26]. One of them is tobermorite, $Ca_5(OH)_2Si_6O_{16}{\cdot}4H_2O$. It is an important material, since it has brilliant physicochemical properties, for example low density 0.5 $g/cm^3$, good ion exchange characteristics. In [27], tobermorite was synthesized by mechanical activation of a mixture of calcium hydroxide and silica in the presence of water at room temperature:

$$5Ca(OH)_2 + 6SiO_2 \rightarrow Ca_5(OH)_2Si_6O_{16}{\cdot}4H_2O.$$

Thermodynamic probability for this reaction to occur is high since the change of Gibbs energy is equal to -303 kJ/mol for the standard conditions.

Mechanical activation of this mixture was carried out in a planetary mill Fritsch Pulverisette - 7. Rotation frequency was 700 rpm, time of treatment was 0.4÷9.5 h. The dependence of tobermorite yield on reagent ratio $CaO:SiO_2$, varied from 1:2 to 3:1, was investigated, as well as the effect of the amount of added water (water/solid ratio varying from 2:1 to 10:1) and time of treatment.

It was stated that weakly crystallized tobermorite was synthesized in a mixture of calcium hydroxide and silica gel. The reaction is most efficient at the reagent ratio $CaO:SiO_2=5/6$ and water-to-solid ratio 4:1. At short activation time, the formation of poor crystallized calcium silicates of B type with tobermorite-like structure was observed. Mechanochemically obtained tobermorite is stable till 800°C. Above this temperature, it decomposes to yield wollastonite. Another form of wollastonite (E), which is a transition phase, is formed from B-type hydrosilicates under heating.

Another example of calcium hydrosilicates with useful properties (incombustibility, high heat-insulating ability, etc.) is afwillite, $Ca_3Si_2O_3(OH)_2 \cdot 2H_2O$. Mechanochemical method of afwillite synthesis was described in [28]. It was stated that afwillite was formed during mechanical activation of calcium hydroxide with silica gel in planetary mill for 2 h in the presence of water added at the amount of 15-30 % mass. At lower and higher water content, the synthesis does not occur. Thus, water content in the mixture under activation plays a decisive role in the mechanochemical synthesis of afwillite. In a mixture activated for less than 90 min, hydrosilicate of B type is formed, as in the case of tobermorite. When heated, it forms the transition E-wollastonite. Afwillite is stable till 700°C; above this temperature it decomposes to give wollastonite $\beta$-$CaSiO_3$.

The preparation of calcium hydrosilicates with Ca/Si = 2 by means of mechanical treatment was investigated in [29]. It is known that it is very difficult to obtain hydrosilicates characterized by the ratio Ca/Si = 2 by any method including hydrothermal synthesis. The mixture under activation was $2CaO + SiO_2$ (silica gel) + 2-fold excess of water (with respect to theoretical amount). Activation resulted in the formation of X-ray amorphous hydrosilicate differed from that prepared by hydrothermal synthesis; it was a mixture of monomers and dimers.

Pure $\beta$-$Ca_2SiO_3$ was obtained as a result of thermal decomposition of this hydrosilicate at temperatures below 1000°C.

**Aluminium silicate (mullite).** Mullite $3Al_2O_3 \cdot 2SiO_2$ is a promising material for modern technology. This is the only compound in $Al_2O_3$ - $SiO_2$ system with melting point at 1850°C. It is characterized by low heat conductivity and thermal linear expansion coefficient, high bending strength, resistance to high-temperature creep; it is also stable to the action of melted metals [30].

It possesses a chain-type structure composed of $AlO_6$ octahedrons and $AlO_4$ and $SiO_4$ tetrahedrons. It belongs to a rhombic syngony with lattice parameters: a = 0.755, b = 0.769, c = 0.2885 nm [31].

The conventional methods of mullite preparation are the following: crystallization from melt, high-temperature sintering of oxides, high-temperature annealing of kaolin [32]. Mechanical activation of hydrated oxides brings to decrease of mullite synthesis temperature [32-39].

The effect of mechanical activation on mullite synthesis from the mixtures of aluminium hydroxide and silica gel was studied in [32,39]. They demonstrated that under heating of nonactivated mixtures, the components undergo phase transformations similar to those observed for individual compounds. Mullite is synthesized at 1400°C.

On the contrary, activated mixture of these compounds exhibits the absence of all effects characteristic of initial components. The observed endothermic effect at 200°C has been assigned to complete dehydration of the mixture. This means that the interaction between components occurs at the stage of activation. $^{27}$Al NMR spectroscopy data confirm this assumption. The joint activation results in the transition of some $Al^{3+}$ cations from octahedral sites into tetrahedral and pentahedral ones (chemical shifts being 52 ppm and 32 ppm, respectively). Note, that coordination of aluminium ions in the initial hydroxide is only octahedral. The authors assume that aluminosilicate structure starts its formation during mechanical activation, and water is formed as a result of condensation:

$$\text{- Si - O - H + H - O - Al -} \rightarrow \text{- Si - O - Al - +} H_2O.$$

The heating of the mentioned mixture leads to the formation of the phase with spinel structure. This phase is transformed into mullite at 1100°C instead of 1400°C in conventional thermal method. Since synthesis temperature is much lower, thus formed mullite is in finely dispersed state ($S_{spec}$~50 $m^2/g$).

A good result was also obtained in the case of the activation of the mixtures of aluminium hydroxide with silicic acid followed by thermal treatment at 1200°C [4]. Practically monophase mullite was obtained (Fig. 6.6).

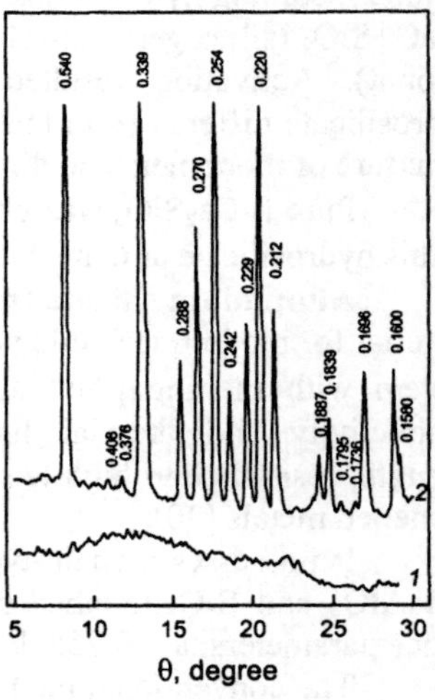


*Figure 6.6.* X-ray patterns of $6Al(OH)_3$+$2SiO_2$·$0.6H_2O$ mixture: 1 – activated, 2 – activated and annealed at 1300°C.

The authors [33-36] carried out comparative investigations of mullite synthesis depending on the nature of initial reagents. The following mixtures were used for synthesis : 1) α-$Al_2O_3$ + silica gel; 2) $Al(OH)_3$ + silica gel; 3) $Al(OH)_3$ + fused $SiO_2$ and 4) $Al(OH)_3$ + silicic acid $SiO_2$·$nH_2O$. The reagents were taken in ratios corresponding to the composition of mullite, $3Al_2O_3$·$2SiO_2$. The mixtures were activated in a planetary mill for 20 h, then studied by means of modern physicochemical methods (DTA, X-ray phase analysis, NMR, IR spectroscopy).

The results obtained were summarized as follows: the best homogeneity of precursors is achieved for the second mixture. Its heating leads to the formation of the spinel phase at ~1000°C, and then mullite is formed at 1200°C. The use of silicic acid decreases the efficiency of mechanochemical synthesis and consequent formation of mullite during heating because of high water content of silicic acid. In the first and third mixtures, the synthesis proceeds at 1400°C, i.e., like in non-activated mixtures.

The authors of [37] confirmed high efficiency of the activation of a mixture of aluminium hydroxide with silicic acid. Monophase mullite was obtained. It was free from iron impurities by treating in weak hydrochloric acid solution. In order to improve agglomeration, the powder, synthesized at 1200°C, was subjected to additional activation for 10 min. Mullite with the density of 3 g/cm$^3$ was obtained.

The authors [38-40] synthesized mullite starting from the other initial components, i.e., mixtures of kaolinite with aluminium hydroxide. According to X-ray data, complete amorphization of the components occurred after activation of the mixture for 60 min. Unactivated mixture provides a 50% yield of mullite at 1200°C, while mixtures activated for 15 min and 120 min provides 80% and 90% yield after heating at 1350°C, respectively. It was found that mullite can be obtained also after high-temperature treatment of mechanically activated kaolinite, $SiO_2$ is formed as an admixture phase [38,39].

**Zirconium silicate (zircon).** Zircon, $ZrSiO_4$, is the main industrial source of zirconium. It is also widely used in production of various types of ceramic materials, such as refractory, glazes, pigments. In this connection, the works on the synthesis of zircon are of great interest [40]. The thermal synthesis of zircon is rather difficult. At low temperatures (1100–1200°C), the process is slow, while at high temperatures (above 1600°C) zircon dissociates into the constituent oxides. Zircon can be obtained, in particular, by sintering within 1350–1550°C range from preliminarily ground and pressed equimolar mixture of anhydrous zirconium and silicon oxides. However, the yield of zircon does not exceed 48 % even at 1550°C. The synthesis temperature can be lowered to 1100°C and zircon yield can be raised to 71–78 % by adding one of the following oxides: 10÷30 % $MoO_3$ (as $Li_2MoO_4$) or ≥2 % $V_2O_5$ [41].

The soft mechanochemical synthesis is a promising method for zircon synthesis. The nature of the starting components is known to strongly affect the parameters of solid phase synthesis. Solid phase synthesis of zircon by 5 min mechanical activation of a mixture of hydrated oxides $ZrO_2 \cdot 3H_2O$ and $SiO_2 \cdot 0.6H_2O$ and subsequent calcination at 1200°C was reported in [4]. Zircon as prepared contained minor amounts of monoclinic and tetragonal modifications of $ZrO_2$ as impurity phases.

Mechanical activation of equimolar mixture of anhydrous and hydrated zirconium and silicon oxides and its effect on solid-phase synthesis of zircon at 1200°C was studied in [42]. The following mixtures were under study: 1) $ZrO_2 + SiO_2$; 2) $ZrO_2 + SiO_2 \cdot 0.48H_2O$; 3) $ZrO_2 \cdot 4.31H_2O + SiO_2$ and 4) $ZrO_2 \cdot 4.31H_2O + SiO_2 \cdot 0.48H_2O$.

Oxide mixtures were mechanically activated in a planetary mill with zirconium jars (jar's volume 0.15 $dm^3$) and balls. The mill rotation frequency was 12.5 $s^{-1}$; energy load was 10 W/g of mixture. The mechanically activated mixtures were thermally treated in air in a VTP-12/15 resistance furnace with lanthanum chromite heaters. Thermal treatment mode was as follows: a rate of temperature rise 300 deg/h, an isothermal exposure at 1200°C for 2 h.

After heating, anhydrous silica contained some amount of water (0.14 mol $H_2O$/1 mol $SiO_2$) and was composed, mainly, of large-sized particles (200–500 μm, 45.52 %). Mechanical activation in the planetary mill for 5 min leads to a dramatic increase in dispersion: powders of both silica and zirconium oxide mainly contain fractions with particle size less than 10 μm. Activated mixtures 1 and 2 have approximately the same granulometric composition, no fractions with particle size in the range 50÷100 μm were observed. Mixture 3 contains fractions with fine particles which predominant in mixtures 1 and 2, as well as fractions with particle size of 50÷100 μm and more, as in the case of mixture 4.

According to XRD, the initial anhydrous zirconium oxide is a monoclinic crystalline modification of $ZrO_2$; anhydrous silicon oxide and hydrated zirconium and silicon oxides are X-ray amorphous substances. Fig. 6.7 represents X-ray patterns for the initial mixture 1, mechanically activated (for 5 min) mixtures 1 and 3, and the same mixtures after thermal treatment. One can see that mechanical activation of a mixture of anhydrous oxide (mixture 1) in the planetary mill for 5 min brings to sharp decrease of the intensity of peaks corresponding to monoclinic zirconium oxide. Mechanical activation of a mixture containing hydrated zirconium oxide (X-ray amorphous) and anhydrous $SiO_2$ (mixture 3) under the same conditions leads to the appearance of low-intensity peaks of monoclinic zirconium oxide.

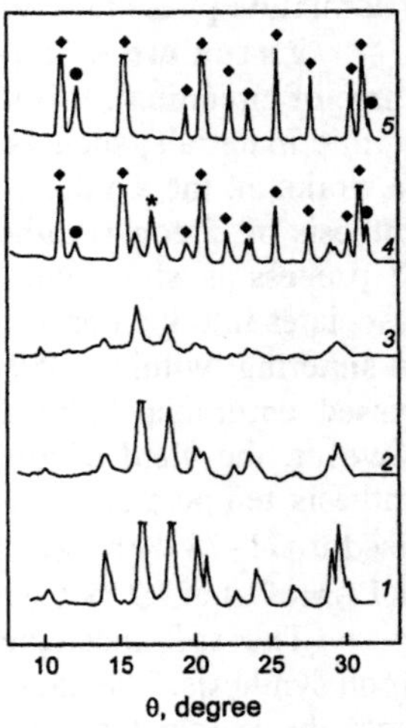


*Figure 6.7.* X-ray patterns of the mixtures of zirconium and silicon oxide:
1 – initial mixture;
2 – mechanically activated mixture (1);
3 - mechanically activated mixture (3);
4, 5 – (2) and (3) after thermal treatment at 1200°C for 2 h. ◆ - $ZrSiO_4$; ★ - $ZrO_2$ (tetragonal); ● - α-$SiO_2$.

X-ray pattern of mixture 1 subjected to thermal treatment contains lines assigned to zircon

$ZrSiO_4$ and low-intensity lines of monoclinic (0.316, 0.284, 0.219, 0.180 nm) and tetragonal (0.295 nm) zirconium dioxide and α-cristobalite (0.405, 0.169 nm). The X-ray pattern of mechanically activated mixture 3 contains lines of zircon and only two low-intensity lines of α-cristobalite. The X-ray pattern of a mechanically activated and thermally treated mixture 2 is practically identical to that of mixture 3, and X-ray pattern of mixture 4 contains weak reflections of monoclinic and tetragonal zirconium dioxide.

The zirconium yield of the synthesis process was estimated by comparing intensities of lines related to one of the main components ($ZrO_2$ and the final product $ZrO_2 \cdot SiO_2$) at the range of 0.345–0.310 nm. This range includes lines of quartz (0.334 nm), monoclinic $ZrO_2$ (0.318 nm), and $ZrSiO_4$ (0.330 nm). According to this technique, the yield of zircon upon thermal treatment of mechanically activated mixtures 2 and 3 is close to 100 %. Thermal treatment of a mechanically activated mixture of anhydrous oxides (mixture 1) gave zircon yield of 89 %. The lowest yield of zircon is observed in the case of mechanical activation of mixture 4 (~35%).

These results show that the nature of reagents subjected to mechanical activation strongly affects the yield of the zircon at subsequent thermal treatment of the mixture at 1200°C. According to X-ray data, the maximum yield of zircon is observed for mechanically activated mixtures containing only one hydrated oxide: either hydrated silica $SiO_2 \cdot 0.48H_2O$ and anhydrous $ZrO_2$ (mixture 2) or hydrated zirconium oxide $ZrO_2 \cdot 4.31H2O$ and anhydrous $SiO_2$ (mixture 3) after their thermal treatment. In the case of the mixture of two anhydrous oxides, the yield of zircon is lower, being the lowest in the case of a mixture of two hydrated oxides (mixture 4). It should be noted that none of the non-activated mixture contains zircon phase upon thermal treatment.

Interaction of the oxides in mixtures 1–4 under mechanical activation was studied by IR spectroscopy, also. The IR spectra of mechanically activated $SiO_2$ differ from those of the initial material: the absorption bands are broadened because of higher disordering of $SiO_4$ polyhedrons compared with the tetrahedral configuration of the initial $SiO_2$. Such polyhedrons in mechanically activated amorphous $SiO_2$ were designed by $A_{Si}$.

To elucidate details of the change in activated anhydrous $SiO_2$ present in activated mixtures 1–4, the spectrum of $SiO_2$ was subtracted from the spectra of the mixtures. Below, only spectra of mixtures 1–4 with subtracted $A_{Si}$ spectrum are discussed. In the IR spectrum of mixture 1 there appears a new low-frequency band ν(Si-O) at 918 $cm^{-1}$ of bridging Si-O-X groups, where X is Si or Zr (Fig. 6.8). The formation of such bridges with weaker Si-O bond indicates that for a part of Si atoms the positions of neighbouring oxygen atoms are changed so strongly that one can speak about the formation of a new oxygen polyhedron (denoted by $B_{Si}$). In the range of Zr-O vibrations (220–650 $cm^{-1}$), four narrow bands of the initial crystalline $ZrO_2$ (577, 500, 415, and 356 $cm^{-1}$) are transformed into broadened bands with changed frequencies. Probably, these bands belong to X-ray amorphous $ZrO_2$, in which the geometry of oxygen coordination polyhedra of zirconium is similar to that of crystalline $ZrO_2$, but differs by greater scattering ability of oxygen atoms and lack of long-range order. Such zirconium polyhedra are denoted as $A_{Zr}$ by analogy with $SiO_4$ polyhedra. The spectra also contain a very broad band at 500 $cm^{-1}$, belonging to strongly distorted oxygen polyhedra of zirconium with strong deviations of oxygen atoms from their optimal positions in crystalline $ZrO_2$. Such polyhedra having no definite geometry are denoted as $B_{Zr}$.

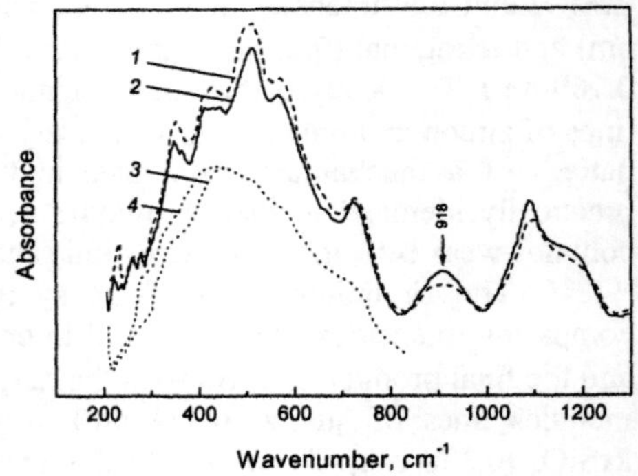


*Figure 6.8.* IR spectra of mechanically activated mixtures I (1), II (2), III (3), IV (4), obtained by subtracting the spectrum of activated $SiO_2$ from the initial spectra.

The spectrum of mixture 2 is similar to that of mixture 2. The only difference is that the peak at 918 $cm^{-1}$, associated with Si-O-X groups, becomes higher, indicating that the fraction of $B_{Si}$ polyhedra increases with respect to $A_{Si}$. The ratio $A_{Zr}/B_{Zr}$ in mixtures 1 and 2 is practically the same.

The spectra of mixtures 3 and 4 differ from those of mixtures 1 and 2. They practically do not contain bands of $B_{Si}$ polyhedra at 918 $cm^{-1}$. Consequently, silicon atoms in mixtures 3 and 4 are surrounded by oxygen polyhedra of $A_{Si}$ type only. The range of the Zr-O frequencies is dominated by a broad band of $B_{Zr}$ polyhedra at ~500 $cm^{-1}$, with the absorption bands of $A_{Zr}$ polyhedra being weak. The spectrum of mixture 4 differs significantly from the spectrum of mixture 3. The peak of the broad absorption band ν(Zr-O) is shifted to lower frequencies by 50 $cm^{-1}$. This means that a part of zirconium atoms in mixture 4 has a different composition of the first coordination sphere: a number of oxygen atoms in it increases. Only $H_2O$ molecules, present in a large excess in mixture 4, can serve as a source of these O atoms.

It should be noted that the bands correspondent to $H_2O$ molecules in mixtures 1–4 is practically the same. The only difference refers to band intensities, which suggests that the content of water in mixtures 1 and 2 subjected to mechanical treatment is the same as in mixtures 3 and 4 exceeding the water content in mixtures 1 and 2 by a factor of 2 and 4, respectively.

The obtained results explain the fact that after thermal treatment of mixtures 1–4 the maximum yield of zircon is observed just in mixtures 2 and 3. It may be assumed that the yield of zircon after calcination is the greater, the higher is the fraction of the amorphous phases $B_{Si}$ and $B_{Zr}$ in mechanically activated mixtures. Indeed, the fraction of the amorphous phase $B_{Si}$ in the mixture 2 is approximately twice higher than that in mixture 1. Presumably, this is due to the replacement of $SiO_2$ by less hard $SiO_2 \cdot 0.48H_2O$. The degree of $ZrO_2$ amorphization and composition of oxygen polyhedra of zirconium remain unchanged.

A similar result was obtained on passing from anhydrous $ZrO_2$ (mixture 1) to less hard hydrated $ZrO_2 \cdot 4.31H_2O$ (mixture 3): the degree of $ZrO_2$ amorphization increases (the fraction of $B_{Zr}$ polyhedra becomes higher), with $SiO_2$ amorphization not reaching the degree corresponding to the formation of $B_{Si}$.

In mixture 4, the extent of $SiO_2$ amorphization is the same as that in mixture 3, but $ZrO_2$ amorphization proceeds differently from that in mixtures 1–3: new $C_{Zr}$ polyhedra with greater number of oxygen atoms are formed in comparable amounts together with $B_{Zr}$ polyhedra. Possibly, oxygen-saturated $C_{Zr}$ polyhedra do not interact with $A_{Si}$ and $B_{Si}$ under thermal treatment of mixture 4, but form a phase of tetragonal $ZrO_2$, and, therefore, the yield of zircon decreases.

The results obtained suggest that the positive role of crystallized water in solid phase synthesis of zircon from a mechanically activated mixture of hydrated oxides consists only in a reduction of the mechanical hardness of the oxides. If one of the two oxides is hydrated, the degree of its amorphization is higher. If both oxides are hydrated, then mechanochemical processes occur, simultaneously with increasing extent of amorphization, to give oxygen-saturated polyhedra of zirconium, which cannot interact with $A_{Si}$ and $B_{Si}$ polyhedra to form zircon in the course of thermal treatment.

Thus, investigations on influence of mechanical activation of equimolar mixture of anhydrous and hydrated zirconium and silicon oxides on the solid phase synthesis of zircon demonstrated that the maximum yield of zircon at 1200°C is achieved only in the mixture containing only one hydrated oxide.

Similar investigations of the mechanochemical synthesis of zircon using anhydrous and sol-gel powders of zirconium and silicon oxides as initial reagents were reported in [43,44].

## 6.3. Titanates

**Magnesium titanate.** A detail investigation of mechanochemical synthesis of magnesium titanate from anhydrous and hydrated magnesium and titanium oxides for preparing the ceramics of the highest quality was carried out in [45]. Four different initial mixtures were selected for the synthesis: 1) $Mg(OH)_2$ + $TiO_2$ (anatase); 2) $Mg(OH)_2$ + $TiO_2$ (sol); 3) MgO + $TiO_2$ (anatase) and 4) $Mg(OH)_2$ + $TiO_2$ (anatase) with excess water to give a slurry. The mixtures were activated in Micros mill for 60 min and then investigated by means of XRD, IR and XPS spectroscopy. According XPS, the formation of Mg–O–Ti bonds was observed only in the first mixture. The activated and unactivated mixtures were calcined then at 900°C for 2 h and sintered at 1350°C for 2 h. After calcination of unactivated mixtures, the reflections of $MgTiO_3$ appeared in the X-ray pattern of mixtures 1-3, but only in the first case the product was monophase. In the mixtures 2 and 3, small amounts of initial oxides were observed. The product formed after heat treatment of the mixture 1 has uniform particles with a size of about 0.3 μm, while in other cases agglomerates of the particles of different sizes were formed. On the other hand, the sintering of activated mixtures 1 and 3 at 1350°C brings to formation of monophase product; in other cases MgO, $Mg_2TiO_4$, $MgTi_2O_5$, $TiO_2$ (rutile), as by-products, were formed. In the mixtures 2 and 4, heterogeneous interaction between $Mg(OH)_2$ particles and $TiO_2$ sol is not efficient because of large amount of water. In the case of anhydrous oxides (mixture 3), magnesium titanate formed as a result of calcination and sintering contains MgO and $MgTi_2O_5$ phases as impurities.

The authors assume that the activation of $Mg(OH)_2$+$TiO_2$ (anatase) mixture involves the transition of proton from OH groups on anatase surface to the hydroxyl group of $Mg(OH)_2$ which results in the formation of water and Mg–O–Ti bond, the latter being a nucleus for the formation of $MgTiO_3$ phase during consequent calcination and agglomeration.

The authors consider that thermal treatment of mixture 1 is accompanied by the growth of nuclei which have been already formed during activation, while in other cases the reaction proceeds via the formation of intermediate products. The sequence of processes for these mixtures is as follows:

1. $Mg(OH)_2 + TiO_2\ (A) \xrightarrow{\text{mech. activation}}$ the formation of Mg–O–Ti bonds at the contacts $\xrightarrow{900°C,\ 2\ h} MgTiO_3 \xrightarrow{1350°C,\ 2\ h} MgTiO_3$ (only grain growth).

2. $Mg(OH)_2\ (sol) + TiO_2\ (sol) \xrightarrow{\text{mech. activation}}$ weak interaction $\xrightarrow{900°C,\ 2\ h} MgO + MgTiO_3 \xrightarrow{1350°C,\ 2\ h} Mg_2TiO_4 + TiO_2(R)$

3. $MgO + TiO_2 \xrightarrow{\text{mech. activation}}$ weak interaction $\xrightarrow{900°C,\ 2\ h} MgO + MgTi_2O_5 + MgTiO_3 \xrightarrow{1350°C,\ 2\ h} MgTiO_3$

4. $Mg(OH)_2 + TiO_2 + H_2O \xrightarrow{\text{mech. activation}}$ no interaction $\xrightarrow{900°C,\ 2\ h} TiO_2$ (A) + $TiO_2$ (R) + $MgTi_2O_5 \xrightarrow{1350°C,\ 2\ h} TiO_2$ (R) + $MgTi_2O_5$

The most perfect microstructure of ceramic is obtained in the case of synthesis in mixture 1 (Fig. 6.9). This structure is formed by tightly packed hexagonal grains. In other cases the structure is less perfect; it is composed of smaller grains; it also contains pores. Therefore, the ceramics as prepared exhibits the best dielectric properties: maximum Q is equal to 9800 at 10 GHz, dielectric constant $\varepsilon$ = 17.3, and $\tau_f$ =-55 ppm/°C.

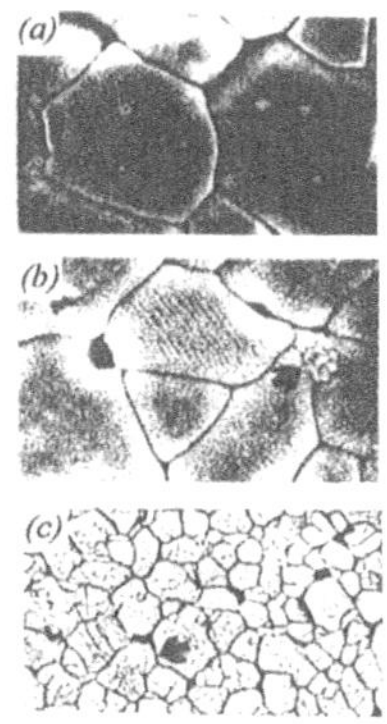

*Figure 6.9.* Microstructure of ceramics polished and the thermally etched at 1250°C for 1 h, prepared from mechanically activated mixtures:
*a* – $Mg(OH)_2$+$TiO_2$(anatase),
*b* - MgO+$TiO_2$(anatase),
*c* - $Mg(OH)_2$+$TiO_2$(sol).

**Calcium titanate**. Calculations of Gibbs energy for the synthesis of calcium titanate, given below, show that from the thermodynamic point of view the reactions involving hydrated titanium dioxide (HTD) are the most profitable:

| | $\Delta G°_{298}$ kJ/mol |
|---|---|
| CaO + $TiO_2$ (anatase) = $CaTiO_3$ | -96.4 |
| CaO + $H_2TiO_3$ = $CaTiO_3$ + $H_2O$ | -164.2 |
| $Ca(OH)_2$ + $TiO_2$ (anatase) = $CaTiO_3$ + $H_2O$ | -40.5 |
| $Ca(OH)_2$ + $H_2TiO_3$ = $CaTiO_3$ + $2H_2O$ | -108.2 |

The authors [46] investigated the effect of mechanical activation on the synthesis of calcium titanate starting from anhydrous and hydrated calcium and titanium oxides.

No common opinion exists concerning the composition and the structure of hydrated titanium dioxide (HTD) used as one of the reagents [47]. Some authors suppose HTD to be a crystal anatase hydrated over the surface. The structure of amorphous HTD differs from that of crystal anatase by the disordering of $TiO_6$ octahedrons due to the presence of a large number of Ti–OH–Ti bridges. According to $^1H$ NMR data [47], HTD contains oxygen-hydrogen groups of three types: 1) three-spin proton configuration ($H_3O^+$, or a proton of $OH^-$ group localized in the vicinity of $H_2O$ molecule); 2) two-spin configuration ($H_2O$ molecule) and 3) relatively isolated hydroxyl groups $OH^-$. Oxygen-hydrogen groups with three-spin configuration are unstable at room temperature and disappear completely after thermal treatment at 110°C. Water molecules are kept in HTD till 240-400°C, while hydroxyl groups till 450-500°C [48]. From the chemical viewpoint, thermal dehydration of HTD is the condensation of ol-bridges and their transformation into oxo-bridges that bind metal ions in oxides. When heated, HTD is dehydrated in two stages, the first one being the removal of water molecules at 60-470°C and the second being the removal of hydroxyl groups at 470-900°C. Dehydration is accompanied by two endothermic peaks with maximum at 170°C and 570°C. Rutile is a final product after heating.

Mechanical activation of HTD is accompanied by its dehydration. The total amount of water in HTD samples activated in planetary mill for 30 min decreases substantially. No noticeable changes in composition and specific surface area are observed during mechanical activation of $Ca(OH)_2$ under the same conditions.

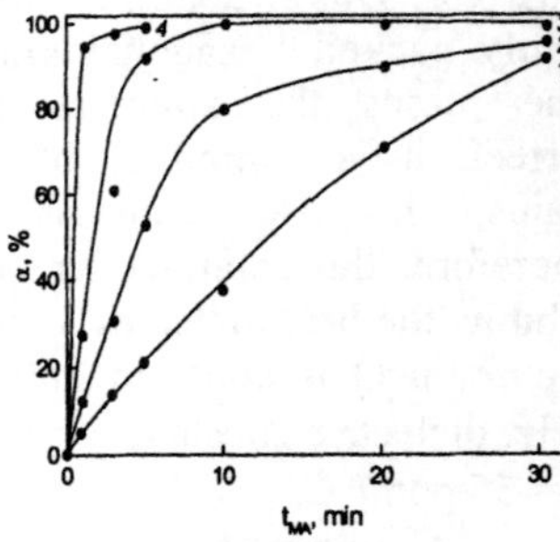


*Figure 6.10.* The change of the interaction degree versus activation time in the mixtures: 1 – $CaO+TiO_2$ (rutile), 2 – $CaO+TiO_2$ (anatase), 3 – $CaO+TiO_2 \cdot 1.2H_2O$, 4 – $Ca(OH)_2+TiO_2 \cdot 1.2H_2O$.

Mechanochemical synthesis was carried out using the following mixtures: 1) $CaO + TiO_2$ (anatase); 2) $CaO + TiO_2$ (rutile); 3) $CaO + TiO_2 \cdot 1.2H_2O$ and 4) $Ca(OH)_2 + TiO_2 \cdot 1.2H_2O$. Molar ratio of reagents was 1:1.

Fig. 6.10 shows the results on free calcium oxide binding in activated above-mentioned mixtures, obtained by means of ethyl glycerine method [46].

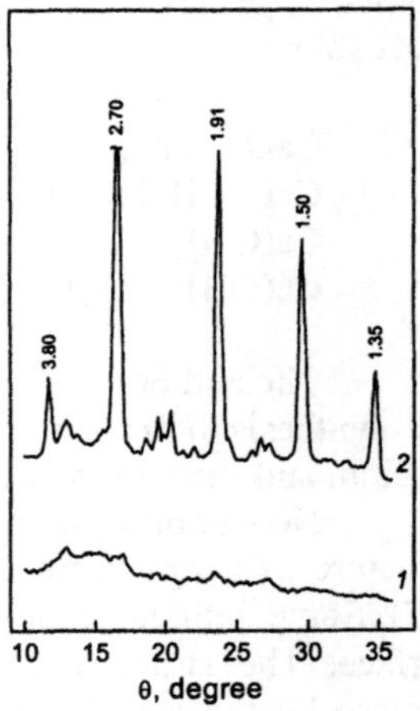


*Figure 6.11.* X-ray patterns of the products of mechanical activation and thermal treatment of $Ca(OH)_2+TiO_2 \cdot 1.2H_2O$ mixture: 1 - 400°C, 2 - 600°C.

As one can see in Fig. 6.11, the binding occurs most fast in the mixtures with HTD; in the mixture of HTD with $Ca(OH)_2$ the product is X-ray amorphous, while in mixture with CaO it is crystalline. Rutile is the least reactive modification, the rate of binding to it being the lowest.

The effect of excess water addition on binding of CaO in $CaO+TiO_2$ (anatase) and $Ca(OH)_2$ +

$TiO_2 \cdot 1.2H_2O$ mixtures was investigated. It was stated that binding rate in the former mixture passed through its maximum corresponding to water content of 5 % mass. For the latter mixture, the maximum was observed at 15 %. It should be noted that in this case the amount of both bound and free water was taken into account; for samples with composition less than 15 %, initial HTD was partially dehydrated at first.

According to IR spectroscopy, hydroxide groups in the structure of the initial $Ca(OH)_2$ are isolated and not bound by hydrogen bonds [49-51]. Their stretching vibrations are responsible for the band at 3630 $cm^{-1}$. The intensity of this band in individually activated $Ca(OH)_2$ samples increases due to the increase of specific surface area. On contrary, proton-containing groups in the initial HTD are characterized by strong hydrogen bond (broad absorption band at 2400–3600 $cm^{-1}$ range). Total decrease of the intensity of this band observed after mechanical activation points to partial mechanochemical dehydration of HTD.

The initial mechanical interaction of $Ca(OH)_2$ with HTD brings to gradual decrease of the intensity of bands corresponding to initial components and to the appearance of an additional shoulder at 2800–3300 $cm^{-1}$ range. Deuteration results in the separation of this band into two ones, at 2620 and 2520 $cm^{-1}$. This indicates that at least two types of OH groups are present in the product (besides the hydroxide groups of $Ca(OH)_2$), differing by the local position and, as a consequence, by binding energy.

Hydroxide groups present on the surface of oxides are known to exhibit either basic or acidic character. One of the methods to determine the surface acidic centres is the sorption of ammonia as a probe molecule. The absence of any noticeable sorption of $NH_3$ on initial and individually activated $Ca(OH)_2$, points out that its surface hydroxide groups are basic ones. On contrary, a part of the surface hydroxide groups of the initial HTD exhibit acidic properties (Bröensted acidic centres), which follows from the formation of $NH_4^+$ ions as a result of proton transfer from the surface Ti–OH groups to $NH_3$ molecule. No substantial changes in the band intensities of the stretching vibrations (2800–3180 $cm^{-1}$) and the bending vibrations of $NH_4^+$ ion occur in activated HTD samples. However, after joint activation of $Ca(OH)_2$ with HTD for 10 s, no noticeable sorption of ammonia on sample surface is observed. Hence, acidic centres of HTD are practically completely consumed at the initial stage via interaction with the surface basic $O^{2-}$ centres (Lewis basic centres) on $Ca(OH)_2$ surface.

$^1$H NMR data confirm the results on the determination of the nature of surface centres. The spectrum of the initial HTD is a superposition of the signals from different proton groups ($OH^-$, $H_2O$ and $H_3O^+$). Mechanical activation of HTD is accompanied by only slight decrease of the number of proton-containing groups of all types. Mechanical activation of $Ca(OH)_2$ leads to an insignificant broadening of the resonance signal due to disordering of its structure. For activated mixture of $Ca(OH)_2$ with HTD, the decrease of total amount of protons is observed pointing to partial removal of water during mechanical activation. The composition of OH groups undergoes substantial changes. Activation brings to sharp decrease of the number of protons corresponding to $H_3O^+$ groups in the mixture. They are practically absent after activation for 10 min. The number of OH groups decreases by nearly 3 times. The fraction of protons in the $Ca(OH)_2$ structure decreases and the contribution from the protons of single and closely located hydroxide groups increases. The latter NMR signals correspond to absorption bands at 2620 and 2520 $cm^{-1}$ in the IR spectra of partially deuterated samples.

Based on data obtained, the authors came to the following conclusion according the mechanism of mechanochemical interaction of $Ca(OH)_2$ with HTD. Due to close contacts between the reagents during mechanical activation, the most mobile acidic protons which are localized at the bridging Ti–O(H)–Ti bonds and form $H_3O^+$-type groups on the surface of HTD diffuse to the surface oxygen atoms of $Ca(OH)_2$ (Lewis basic centres). As a consequence, the formation of water molecules and Ca–O–Ti bonds occurs (condensation mechanism). The ratio between the consumed acidic (HTD) and basic ($Ca(OH)_2$) centres should be equal to 1:1. However, this ratio was found to be higher. Probably, coordination-unsaturated titanium atoms arising during activation coordinate either oxygen atoms of isolated Ti–OH groups, transforming them into bridge groups, or oxygen atoms of water molecules. Thus, a constant generation of acidic protons occurs. This process is accompanied by the destruction of HTD crystal structure starting from the surface. Strong bridge oxo-bonds Ti-O-Ti turn into weak ol-type bonds Ti-OH...Ti.

The following scheme of this process was proposed in [50]:

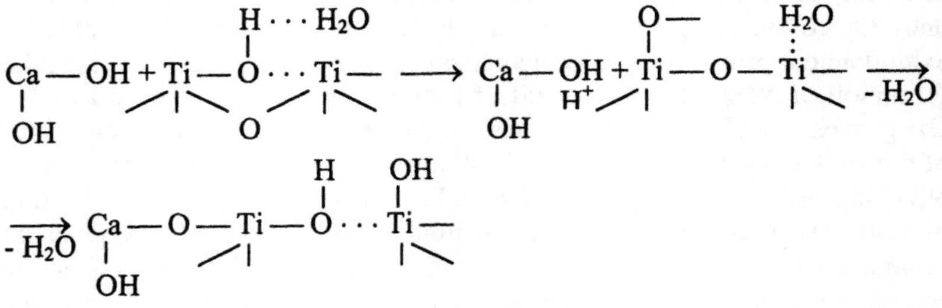


The transport of protons from HTD surface on $Ca(OH)_2$ surface is accompanied by redistribution of electron densities on neighboring calcium and titanium atoms [50]. According to XPS (Fig. 6.12), the binding energy of the Ti $2p_{3/2}$ level in the initial HTD is weakly sensitive to the presence of $OH^-$ groups and $H_2O$: it is quite the same for HTD and anhydrous anatase [51]. In the activated HTD, the position of the maximum of Ti $2p_{3/2}$ line is not changed but broadening of the line occurs that is connected with the formation of defects on the surface. However, in the spectrum of activated mixture of HTD with $Ca(OH)_2$, binding energy of Ti $2p_{3/2}$ level is decreased by 0.7 eV. This energy corresponds to the binding energy in the final product, $CaTiO_3$, which is formed after calcination. To authors opinion, a decrease of binding energy of Ti $2p_{3/2}$ level for activated mixtures is a consequence of the formation of Ca-O-Ti bonds that is confirmed by NMR.

Mechanical activation of $Ca(OH)_2$ is not accompanied by the change in the position of Ca $2p_{3/2}$ line. For activated mixture of $Ca(OH)_2$ with HTD, its maximum is shifted, and the bond energy increases by 0.5 eV with respect to the final product, $CaTiO_3$. After annealing of activated mixture at 1000°C and the formation of $CaTiO_3$, a positive shift of Ca $2p_{3/2}$ line disappears.

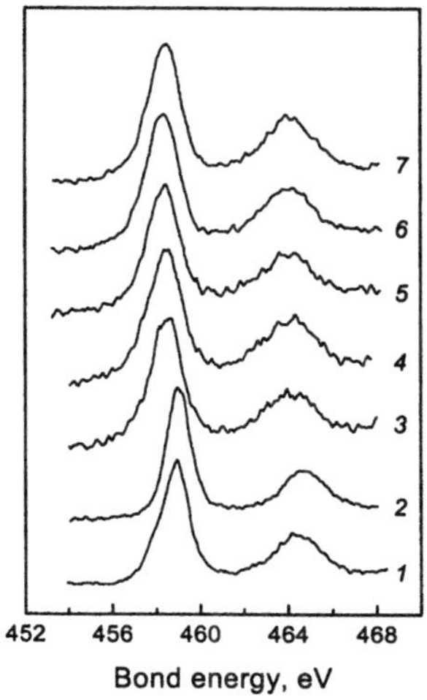


*Figure 6.12.* X-ray electron spectra of the $Ti_{2p}$ level:
1 - $TiO_2$ (anatase); 2 – $TiO_2{\cdot}1.2H_2O$;
3-6 – $Ca(OH)_2$+$TiO_2{\cdot}1.2H_2O$ mixtures activated 1, 2, 5 and 10 min; 7 – sample 6 after annealing at 1000°C.

The largest changes are observed in the spectra of O 1s level. The components connected with oxygen atoms of the oxide group (530.2 eV) and oxygen atoms of proton-containing groups (531–532 eV) are observed in the spectra of initial HTD. Bond energy of the O 1s level in the initial $Ca(OH)_2$ differs from that of HTD and is equal to 530.9 eV. At short activation time, the spectrum of O 1s is a superposition of the main component at 531.3 eV and a shoulder around 530 eV. Later on, their intensities become equal. After annealing at 1000°C, the component at 529.4 eV becomes the main one with low-intensity shoulder at 531.3 eV. This change of spectra is explained by the fact that oxygen atoms of proton-containing groups are predominant in the surface layer of the initial mixture but their amount in activated samples decreases due to removal of water from the surface.

The calculation of the changes of Ca, Ti, and O atom concentrations, carried out on the basis of the measurements of intensities of XPS lines, showed that Ca and O atoms are present in excess (with respect to stoichiometry of $CaTiO_3$) in the surface layer of activated mixtures (Fig. 6.13). With increasing time of mechanical activation, their content decreases monotonously till the level corresponding to $CaTiO_3$ (O/Ti = 3, Ca/Ti = 1). This fact is the evidence that the surface of the particles being formed is covered with $Ca(OH)_2$ layer. Further activation brings to decrease of Ca atoms on the surface as a result of intensive diffusion and the formation of Ca-O-Ti bonds. At the same time, a decrease of the number of oxygen atoms incorporated into proton-containing groups is observed due to the removal water molecules from the surface.

Thus, it was stated that the initial stages of mechanochemical reaction of $Ca(OH)_2$ with HTD are connected with the transport of mobile protons from the acidic oxide to basic one, and with the change of electron density at the neighboring titanium and calcium atoms. These processes result in the formation of water molecules and Ca-O-Ti bonds.

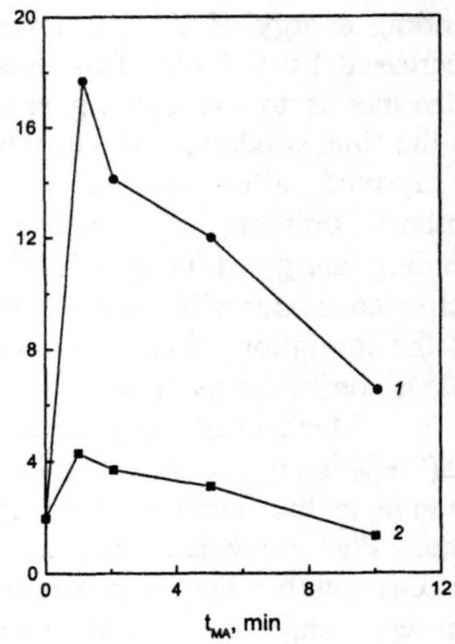


*Figure 6.13.* Relative changes in the concentrations of elements in the surface layer of $Ca(OH)_2+TiO_2 \cdot 1.2H_2O$ activated mixtures, according to XPS data: 1 – O/Ti; 2 – Ca/Ti.

At the next stages with higher transformation degree, the formation of intermediate products is possible, in particular calcium hydrotitanates. As it was demonstrated in [52], the compound $CaO \cdot 4TiO_2 \cdot 6H_2O$ is formed as an intermediate product. Its existence was proved by means of the method of differential dissolution of mechanically activated mixtures of calcium hydroxide with metatitanium acid and by X-ray method.

**Barium titanate.** The synthesis of barium titanate on the basis of HTD and barium dioxide was reported in [53]. The following reactions were studied:

| | $\Delta H^\circ_{298}$, kJ/mol |
|---|---|
| $BaO_2 + TiO_2 = BaTiO_3 + 1/2O_2$ | -92.21 |
| $BaO_2 + H_2TiO_3 = BaTiO_3 + H_2O + 1/2O_2$ | -166.36 |

The mixtures were activated in a planetary mill AGO-2 in air. The mass ratio of materials to balls was 1:20, activation time was 1-5 min.

Fig. 6.14 shows X-ray diffraction patterns of the products obtained after activation of the mixtures of $BaO_2$ with rutile, anatase and metatitanium acid. It follows, that mechanochemical synthesis of barium titanate occurs only if anatase and metatitanium acid are used as initial reagents. The presence of barium carbonate phase was observed for all the products. For the mixture with anatase, the amount of barium carbonate increases with activation time, and for the mixture with metatitanium acid it decreases. In the latter case, the product of interaction, $BaTiO_3$, is well crystallized.

The synthesis from HTD and barium oxide was also carried out under the conditions of shock compression [54]. The yield of $BaTiO_3$ was 70%.

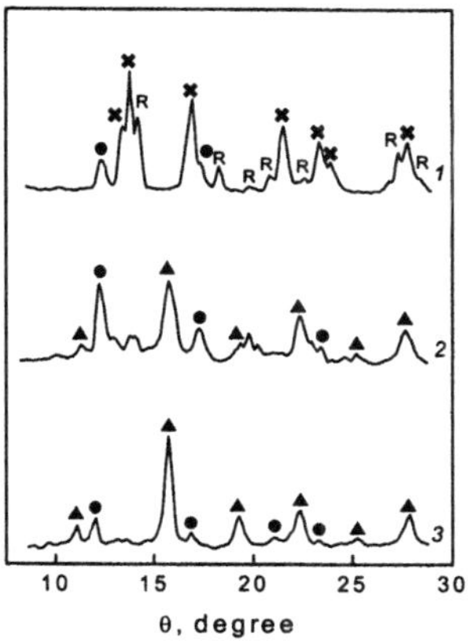


*Figure 6.14.* X-ray patternes of activated mixtures of $BaO_2$ with rutile (1), anatase (2), and metatitanic acid (3). Activation time - 5 min.
▲ - $BaTiO_3$, ✖ - $BaO_2$, ● - $BaCO_3$, R – $TiO_2$, rutile.

**Aluminium titanate.** Ceramic based on aluminium titanate is widely used in technology and industry because it is characterized by small thermal expansion coefficient and high melting point [55]. Usually aluminium titanate is obtained by long agglomeration of a mixture of aluminium and titanium oxide powders. It was noted that, despite the diversity of preparation methods, aluminium titanate is formed only when the mixture is annealed at a temperature above 1400°C [56]. Aluminium titanate is slowly decomposed to give the initial oxides during long heating and slow cooling within the range 750-1200°C.

Aluminium titanate was obtained by the annealing of the activated mixture of aluminium hydroxide and metatitanic acid at a temperature of 1300°C (Fig. 6.15) [4]. The authors of [57,58] consider that mechanical activation of aluminium and titanium hydrogels is accompanied by the formation of Al-O-Ti bonds, while the heating of this mixture gives aluminium titanate. It is stressed that the efficiency of mechanical activation is achieved not only due to the contacts between dispersed particles but also due to electron redistribution leading to the formation of a new chemical bond. It was demonstrated in [59] that the addition of talc ($3MgO{\cdot}4SiO_2{\cdot}H_2O$) has a positive effect on the synthesis of aluminium titanate by means of the activation of a mixture of aluminium hydroxide and metatitanic acid followed by annealing.

A detailed investigation of soft mechanochemical synthesis of aluminium titanate was described in [60]. The data obtained in this study are the evidence of the formation of amorphous hydrated aluminium titanium compounds at the initial stages. These compounds are stable till 450-700°C; at higher temperature they decompose to give aluminium and titanium oxides which form dispersed aluminium titanate at 1340°C.

An interesting way of aluminium titanate preparing was proposed in [61,62]. A mixture of metallic aluminium with $TiO_2{\cdot}H_2O$ hydrogel was activated mechanically in a ring roll mill. Aluminium was oxidized and $Ti^{4+}$ was partially reduced to $Ti^{3+}$ during the activation of the mixture for 300 min. Heating of this mixture at high temperature leads to the synthesis of $Al_2TiO_5$.

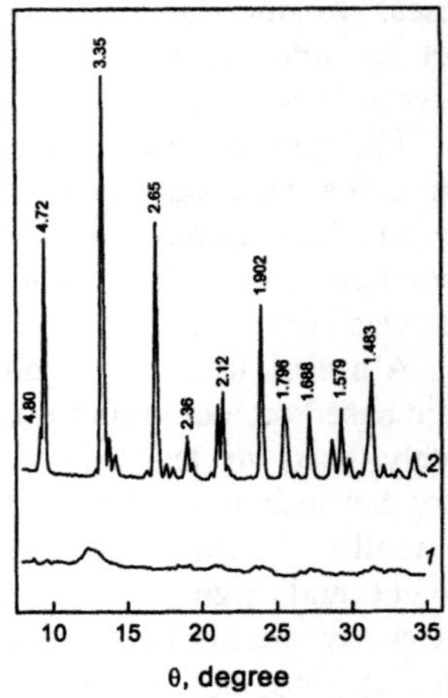


*Figure 6.15.* X-ray patterns of activation product (1) and the product of subsequent annealing at 1300°C (2) of the mixture $2Al(OH)_3$+$TiO_2{\cdot}H_2O$.

**Titanates of rare earths.** The authors of [63] investigated the formation of dititanates $Ln_2Ti_2O_7$ (Ln = La, Ho, Dy, Gd, etc.) after mechanical activation of $Ln_2O_5$ + $2TiO_2$ and $Ln(OH)_3$ + $2Ti(OH)_4$ mixtures.

The mixtures were treated in vibratory mill at different load parameters. Mechanical activation caused the decrease of synthesis temperature from 1200°C (calcination for 100 h) to 850°C (calcination for 20 h) in both cases. An increase of the rate is determined by the decrease of particle size, the existence of good contacts between them, and disordering of the structure of initial components.

Mechanical activation of hydroxides at room temperature leads to the formation of a complex compound $Ln_2[TiO_2(OH)_2]_2{\cdot}nH_2O$. This complex is stable when heated up to 550-600°C. Above 600°C, it decomposes to form $Ln_2Ti_2O_7$.

**Zirconium titanates.** In $ZrO_2$ – $TiO_2$ system, the compound $Zr_2TiO_4$ is formed at 1000°C. Its structure belongs to orthorhombic syngony with the parameters a = 0.4804, b = 0.5482, c = 0.5091 nm. If the reagents ratio is slightly changed, $Zr_5Ti_7O_{24}$ is formed; this compound forms solid solutions with $Zr_2TiO_4$. Investigations showed that $Zr_5Ti_7O_{24}$ possesses an ordered superstructure of α-$PbO_2$ type with lattice parameters a = 1.4357, b = 0.5325, c = 0.5020 nm [64].

The known methods of $Zr_5Ti_7O_{24}$ synthesis include long high-temperature treatment of a mixture of zirconium and titanium oxides at 1500°C for 4 h in air.

Hydrated oxides $ZrO_2{\cdot}nH_2O$ and $TiO_2{\cdot}H_2O$ were used as initial reagents for the synthesis of zirconium titanates. After the joint activation of the mixture for 15 min in a planetary mill, the mixture was heated at 1260°C for 4 h. These procedures resulted in the formation of both $Zr_2TiO_4$ and $Zr_5Ti_7O_{24}$. X-ray diffraction patterns of compounds as prepared are shown in Fig. 6.16. The reflections completely correspond to those reported in [65].

The compound $Zr_5Ti_7O_{24}$ has a mixed structure with alternating fragments of the structures of the two end members of the row. This ordering removes the excess mechanical strain in the lattice and causes the decrease of dielectric losses. This compound is used in high-frequency technology. $Zr_5Ti_7O_{24}$ as prepared is characterized by the following parameters: $\varepsilon = 34$–2, Q = 660–700 at the frequency of 1 kHz [66].

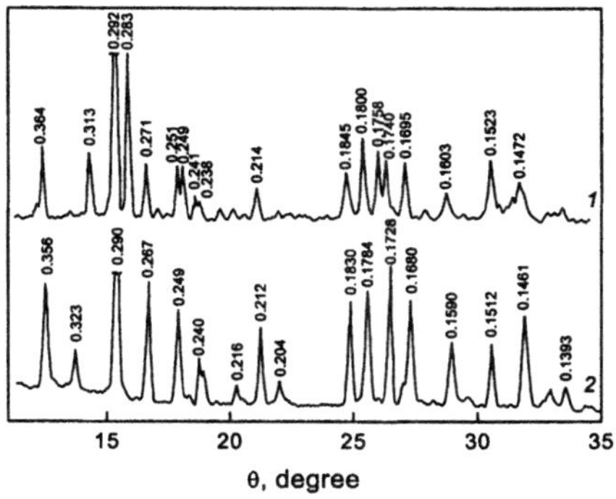


*Figure 6.16.* X-ray patterns of zirconium titanates synthesized by mechanical activation and subsequent annealing at 1260°C for 4 h of $ZrO_2 \cdot nH_2O + TiO_2 \cdot H_2O$ mixtures.
1- $ZrTiO_4$, 2 – $Zr_5Ti_7O_{24}$.

## 6.4. Vanadates

Most of oxide vanadium compounds of alkaline metals have been related to vanadium bronzes due to similarity of their properties to tungsten bronzes. They are widely used in metallurgy, catalysis, electrochemistry and other fields of industry [67]. As a rule, these compounds are prepared by sintering of $V_2O_5$ with the salts of corresponding metals at high temperatures, that brings to formation of poison gases. The other possible way to prepare these materials is a low-temperature method which is realized by coprecipitation of corresponding compounds and the following drying of final product at 300-400°C. The main disadvantage of such method consists in wastes formation. In this connection, mechanical activation appeared to be a prospective method for the synthesis of vanadates.

**Sodium and potassium vanadates.** The authors [68] investigated mechanochemical synthesis of sodium and potassium decavanadates starting from $V_2O_5$ and corresponding carbonates or metavanadates. They have shown that varying the initial ratio of components and the time of mechanical activation of the mixtures, the synthesis of different vanadates, including decavanadates, occurs, characterized by high rate of reactions.

**Lithium vanadium oxide $LiV_3O_8$.** In $Li_2O$-$V_2O_5$ system within the range 50-100% $Li_2O$, two $V^{5+}$-containing compounds exist: $LiVO_3$ and

$Li_3VO_4$ [69,70]. Vanadium compounds formed in air within the range 0-50% $Li_2O$ are attributed to lithium-vanadium bronzes. $Li_{1+x}V_3O_8$ is a bronze-type compound in which a part of $V^{5+}$ ions are reduced to $V^{4+}$. $Li_{1+x}V_3O_8$ crystallizes in monoclinic system. The structure of $Li_{1+x}V_3O_8$ is closely related to those of $V_6O_{13}$. It consists of double and single zigzag octahedral strings arranged so as to provide interlayer sites (tetrahedral and octahedral) for lithium ions. It is achieved by expanding framework of vanadium polyhedrons.

Last years this compound is widely studied as a cathode material for rechargeable lithium batteries because of its high capacity, facile preparation and stability in air [71-73]. $Li_{1+x}V_3O_8$ is usually prepared either by high temperature methods (e.g., by melting of a mixture of $Li_2CO_3$ and $V_2O_5$ at 680°C) [71] or by low-temperature methods (e.g., by co-dissolution of lithium and vanadium compounds and the following drying of the product at 300-400°C) [72,73]. It was shown that solution technique, producing more disperse or even amorphous material, should bring to higher capacity and better cycling behavior due to decreasing of particles' size, a degree of crystallinity and an absence of the long-range order, reducing the pathway through lithium ions diffusion.

Mechanochemical synthesis of $Li_{1+x}V_3O_8$ starting from $V_2O_5$ and LiOH was studied in [74-76] by X-ray, IR and EPR spectrscopy. The $LiOH+V_2O_5$ mixtures were activated in high energetic planetary mill with corundum jars and balls.

It was shown (Fig. 6.17) that intensity of reflections of $V_2O_5$ on X-ray patterns of activated mixtures decreases versus time of activation, while the reflections of initial LiOH from the very beginning are present at the background level due to its amorphization and weak scattering ability of lithium atoms. On the pattern of the sample activated for 1 min, one can see the appearance of new weak reflections assigned to lithium-vanadium bronzes. After mechanical activation for 10 min, the new reflections are corresponding to $Li_{1+x}V_3O_8$. It indicates that structural transformations in the mixture of

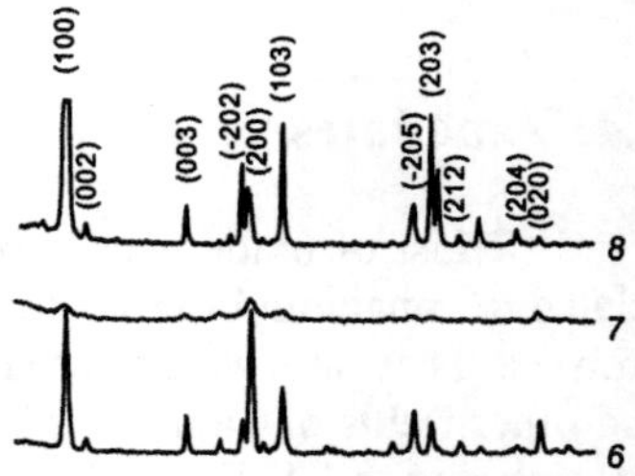


*Figure 6.17.* X-Ray patterns of LiOH + $V_2O_5$ mixture, activated 30 sec (2); 1 min (3); 5 min (4); 10 min (5); 10 min activated and heated at 400°C (6); 5 min activated and aged (7) in comparison with initial $V_2O_5$ (1) and HT-$LiV_3O_8$ (8).

solid reagents occur at the early stages of mechanical activation.

In IR spectra of activated samples, the intensity of absorption band at 1020 $cm^{-1}$, corresponding to stretching vibrations of double vanadyl bonds ν (V=O) in $V_2O_5$, decreases (Fig. 6.18). New bands appear at 900-1000 $cm^{-1}$ range: 920, 950 and 995 $cm^{-1}$, which could be also assigned, according to data [77], to V=O stretching vibrations. The change of vibration frequency of V=O bonds in activated samples as well as in final product, $Li_{1+x}V_3O_8$, in comparison with pure $V_2O_5$, is a result of incorporation of lithium ions into the cell.

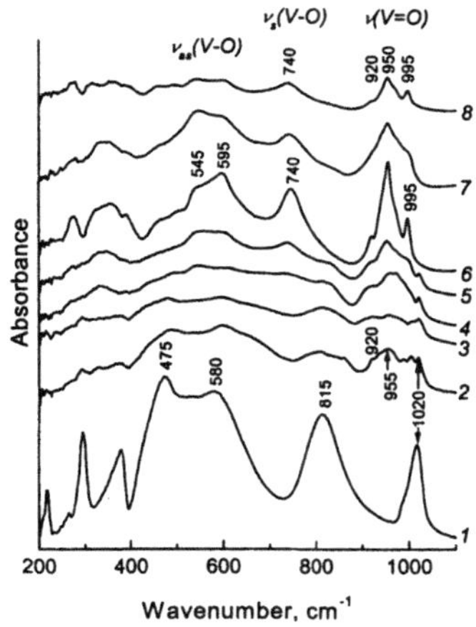


*Figure 6.18.* IR spectra of LiOH + $V_2O_5$ mixture, activated 30 sec (2); 1 min (3); 5 min (4); 10 min (5); 10 min activated and heated at 400°C (6); 5 min activated and aged (7) in comparison with initial $V_2O_5$ (1) and HT-$LiV_3O_8$ (8).

The phase composition of activated samples is significantly changed after ageing for 6 months at room temperature. The intensity of X-ray reflections of $V_2O_5$ is sharply decreased, while those of $Li_{1+x}V_3O_8$ is increased. In IR spectra the increase of bands' intensity is observed. The complex line at 900-1050 $cm^{-1}$ range become narrower, clear maximum at 950 $cm^{-1}$ with a shoulder at 970 $cm^{-1}$ appears.

After heating of activated mixtures at 400°C, all reflections observed belong to $Li_{1+x}V_3O_8$. Note, that heating of nonactivated mixtures at 400°C (that is lower than the melting temperature of LiOH) for 4 h brings to partial interaction of reagents only. Thus, the preliminary mechanical activation, even for a short time (30 sec), essentially decreases the final temperature of $Li_{1+x}V_3O_8$ synthesis, providing formation of product in high disperse state.

EPR spectroscopy data indicate that structural transformations observed in the activated LiOH+$V_2O_5$ mixture as well as after its heating or ageing are accompanied by reduction processes.

Mechanical activation of $V_2O_5$+LiOH mixture brings to the change of EPR spectra as time of activation increases (Fig. 6.19). The spectra observed for the mixtures activated for 1-5 min can be described as a superposition of two signals. The parameters of the first signal ($g_{\perp}$=1.98; $g_{\parallel}$=1.92; $A_{\perp}$=65 G; $A_{\parallel}$=185 G), which is more clearly displayed after 5 min, well correspond to parameters of EPR spectra of isolated vanadyl ions, $V=O^{2+}$, observed for mechanically activated $V_2O_5$ [78] and lithium-doped $V_2O_5$ monocrystals [79]. Affiliation of this signal to activated $V_2O_5$ was confirmed by the following observation. It was found that after half-a-year ageing at room temperature, this signal practically disappears in the spectrum of 5 min activated mixture due to consumption of $V_2O_5$ in reaction with LiOH.

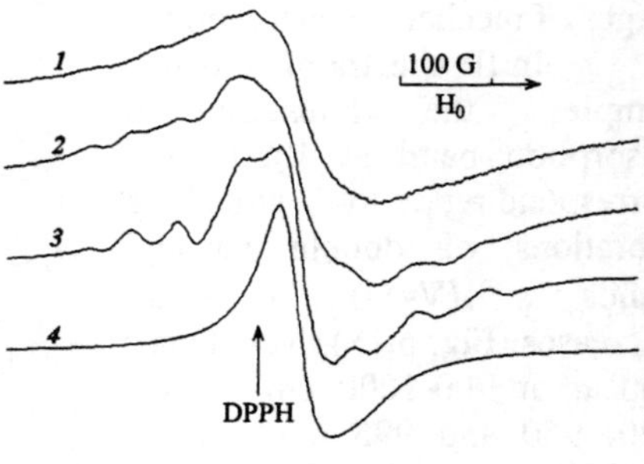

*Figure 6.19.* EPR spectra of LiOH+$V_2O_5$ mixture, activated 30 sec (1), 1 min (2), 5 min (3) and 10 min (4), measured at 77 K.

The spectra of 10 min activated mixture and those activated 5 min and aged are similar to each other and can be practically described by second signal only with $g_\perp$=1.96 and weakly anisotropic form (sloping at $g_{||}$ range), and without traces of hyperfine structure (HFS) and fine structure. The parameters and the form of this signal is completely correspondent to those of HT-$Li_{1+x}V_3O_8$. Heating of 10 min activated samples at 400°C doesn't bring to significant change of EPR spectra. A rather big $g_\perp$ value of signal observed allows to assign it to $VO^{2+}$ ions with vanadyl bond as it is a characteristic feature of these ions [79].

The absence of HFS in the EPR spectrum of 10 min activated mixture points to the appearance of weak exchange interaction between $V^{4+}$ ions (localized centres), probably, through electron gas (delocalized electrons) - C-S-C relaxation [80,81].

It is known that a direct consequence of the presence of electron gas in oxides is an observation of temperature dependence of linewidth of EPR spectra of paramagnetic centres. For centres located in crystal fields of low symmetry, e.g., $VO^{2+}$ ion with short vanadyl bond, the linewidth is independent of temperature. The presence of temperature dependence in this case can be considered as a direct evidence of the appearance of electron gas [78]. Table 1 shows that for 5 and 10 min activated samples, the linewidth of spectra representing only anisotropic signal is changed as measurement temperature increases from 77 to 293 K.

Uncommon peculiarity of EPR spectra of $Li_{1+x}V_3O_8$ obtained by heating of activated mixtures, as well as of HT-$Li_{1+x}V_3O_8$, consists in the sloping form of anisotropic signal at $g_{||}$ range without clear peak corresponded to $g_{||}$. Such spectrum can be assigned to the system of exchanged localized $V=O^{2+}$ centres of two types, differing by direction of a crystal field axe of a short V=O bond. Probably, the reason of the change of crystal field axe of about 50% of $V=O^{2+}$ ions is the presence of $Li^+$ ions in the first coordination sphere in vicinity of vanadyl oxygen.

Thus, mechanical activation of $V_2O_5$+LiOH mixtures combined with low temperature treatment brings to formation of reduced $Li_{1+x}V_3O_8$ phase with two types of $V=O^{2+}$ ions (localized electron centers) and a lot of conductive electrons (electron gas).

Chemical analysis shows that the content of $V^{4+}$ ions in $Li_{1+x}V_3O_8$, prepared by heating of activated mixtures at 400°C, is of the same order as in HT-$Li_{1+x}V_3O_8$, while specific surface area values of $Li_{1+x}V_3O_8$ as prepared is 4 times higher than of HT-$Li_{1+x}V_3O_8$ and is equal to ~ 7 $m^2$/g.

**Calcium vanadates.** In CaO-$V_2O_5$ system the existence of three $V^{5+}$ containing compounds is stated: metavanadate, $Ca(VO_3)_2$, pirovanadate, $Ca_2V_2O_7$, and ortovanadate, $Ca_3(VO_4)_2$. The interaction of $V_2O_5$ with different calcium compounds under heating was investigated by numerous researches. The authors [67, 82] showed that in CaO-$V_2O_5$ mixture the formation of initial product - metavanadate $CaV_2O_6$ - occurs noticeably above 500-520°C in solid state. The sequence of phase transformations, contrary to thermodynamic estimations, is correspondent to the increase of calcium content in complex oxide [82]. It was noticed also, that the rate of calcium vanadate formation is higher if $Ca(OH)_2$ is used instead of anhydrous CaO, most probably, due to the conjunction of the interaction process and $Ca(OH)_2$ decomposition [82].

The authors [83] have revealed the increase of reactivity of crystalline $V_2O_5$ in its equimolecular mixture with calcite which was treated at 150°C in a saturated water-vapour atmosphere. After 192 h, the formation of crystalline calcium metavanadate was observed. The occurrence of the reaction was explained by decrystallization of initial crystalline $V_2O_5$ during the process of adsorbtion of water molecules with the formation of very acid reactive gels in the presence of basic compounds ($CaCO_3$). The following explanation of action of water on amorphous $V_2O_5$ was proposed: the polar water molecules are absorbed on the surface of the polymeric V-O chains, progressively breaking the three-dimensional network of amorphous $V_2O_5$ that gives rise to strongly one-dimensional $V_2O_5$ gels of layered structure with basal d-spacing dependent on the interfoliar water content. The acid catalytic properties of these gels seem to be evident from the $CaCO_3$ decomposition at 150°C and posterior $CaV_2O_6$ formation. I.e., the adsorption of $H_2O$ molecules increases the reactivity of the system through the formation of active acid sites.

Mechanochemical synthesis of calcium vanadates, starting from anhydrous and hydrated calcium vanadates, was studied in [84]. The atomic ratio Ca/V in the mixtures of CaO+$V_2O_5$, CaO+$V_2O_5\cdot H_2O$, $Ca(OH)_2$+$V_2O_5$ and $Ca(OH)_2$+$V_2O_5\cdot H_2O$ was correspondent to meta-, piro- and ortovanadates. According to XRD (Fig. 6.20), mechanical activation brings

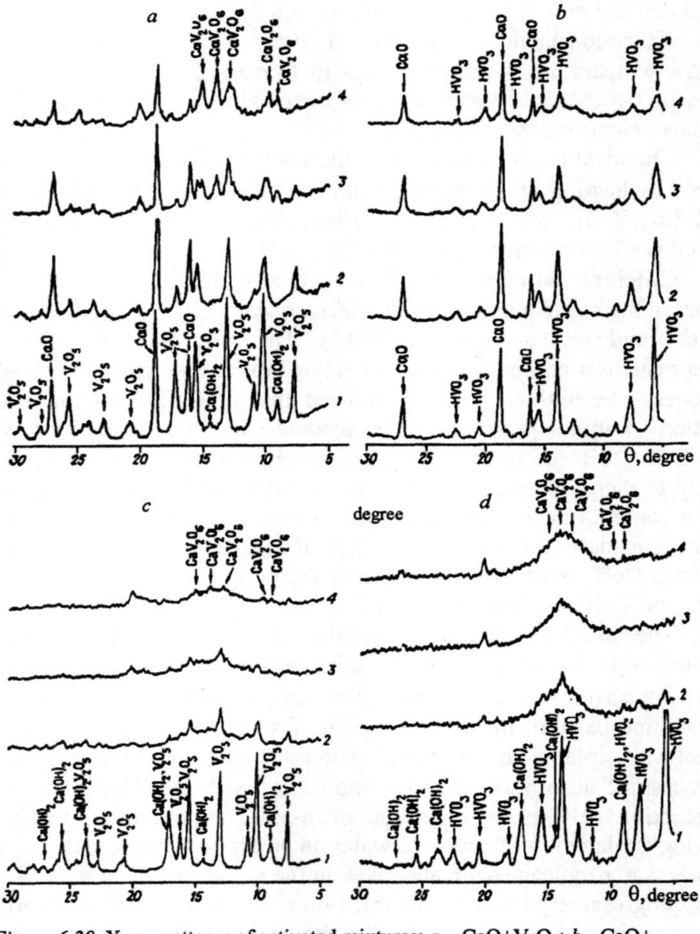


*Figure 6.20.* X-ray patterns of activated mixtures: *a* – $CaO+V_2O_5$; *b* - $CaO+V_2O_5{\cdot}H_2O$; *c* – $Ca(OH)_2+V_2O_5$; *d* – $Ca(OH)_2+V_2O_5{\cdot}H_2O$.
Time of activation: 1 (1); 5 (2); 10 (3); 15 (4) min.

to 3 different results depending upon initial reagents being used: 1) the formation of crystalline final product $CaV_2O_6$ in the mixture of anhydrous oxides; 2) the absence of noticeable interaction in $CaO+V_2O_5{\cdot}H_2O$ mixture; 3) the formation of amorphous product in $Ca(OH)_2+V_2O_5{\cdot}H_2O$ and $Ca(OH)_2+V_2O_5$ mixtures. The same results were observed in the mixtures with Ca/V ratio equal to 2/1 and 3/1. The increasing of activation time brings to crystallization of the product, while the higher is the amount of $Ca(OH)_2$ in the initial mixture and $H_2O$, respectively, the higher is the rate of the process. The phase composition of products is changed simultaneously: the formation of vanadates enriched with calcium becomes preferable.

X-Ray amorphous products obtained in the result of mechanical activation of $Ca(OH)_2+V_2O_5$ and $Ca(OH)_2+V_2O_5{\cdot}H_2O$ mixtures were studied by IR spectroscopy. In IR spectra of samples as prepared, the band correspondent to stretching vibrations of vanadyl bond V=O in the spectra of initial $V_2O_5$ and $V_2O_5{\cdot}H_2O$ ($\nu$=1020 $cm^{-1}$) is shifted to low-frequency range (Fig. 6.21). The shift is determined by the appearance of calcium ions in the structure in the vicinity of vanadium ions with vanadyl bond, as in the case of $LiV_3O_8$ (see above). In the spectra of activated mixtures one cannot observe the vibration band of isolated OH-groups, present in the spectrum of initial $Ca(OH)_2$. Instead of it, the wide band appears at 2700-3600 $cm^{-1}$ range pointing to the presence of proton containing groups with strong hydrogen bonds.

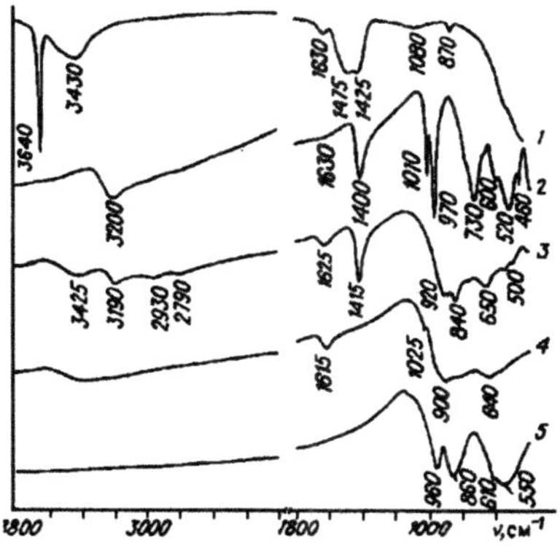

*Figure 6.21.* IR spectra of initial compounds (1, 2), activated mixtures (3, 4) and final product (5):
1 – $Ca(OH)_2$; 2 - $V_2O_5{\cdot}H_2O$;
3 – $Ca(OH)_2+V_2O_5{\cdot}H_2O$;
4 - $Ca(OH)_2+V_2O_5$; 5 – $CaV_2O_6$.

The formation of amorphous phase in the course of mechanical activation of mixtures of acids and bases in high energetic activators without cooling of jars (i.e., without the removal of heat) speculates that water molecules appeared under mechanochemical interaction transform into the structural component of intermediate amorphous phase (matrix) and stabilize it.

According to thermal analysis, the activated samples have some amount of $V^{4+}$ ions. Their oxidation occurs at temperature above 450°C. The highest amount of reduced ions is observed in $Ca(OH)_2+V_2O_5$ mixture.

Heat treatment of activated samples prepared from hydrated oxides at 500°C for 2 h yields the single phase crystal product – $CaV_2O_6$. Note, that in the case of nonactivated mixtures of anhydrous or hydrated oxides the same treatment brings to partial interaction only.

**Strontium vanadates.** Authors [85] studied mechanochemical synthesis of $SrVO_3$ with perovskite structure. Initial mixtures comprising $Sr(OH)_2$ with $V_2O_5$, $VO_2$ or $V_2O_3$ were milled in a multi-ring type mill (MICROS MIC-0) at 2000 rpm for 60 min in air or $N_2$-4% $H_2$ and then heated at 800°C for 10 h. Single phase $SrVO_3$ was obtained by firing of milled mixtures at 1300°C under reducing atmosphere with 4 dehydration steps via two intermediates, $Sr_2V_2O_7$ and $Sr_8V_2O_{2.5}$. Dehydration of the mixtures was promoted by milling under $N_2$ – 4% $H_2$, due to reduction of precursors by hydrogen. The precursor from the mixture with $V_2O_3$-$V_2O_5$ milled under $N_2$ – 4% $H_2$ appeared to be closer to the final product, $SrVO_3$, than the other precursors.

**Vanadates of p- and d- elements.** Vanadates of p- and d- elements as well as $V_2O_5$ supported on their oxides are widely used as catalysts of selective oxidation of different organic compounds and selective catalytic reduction of $NO_x$ by ammonia. As a rule, such catalysts are prepared by heating of the mixture of corresponding oxides or ammonium salts. High temperatures of the processes bring to high-energy consumption and elimination of poison vapours of vanadium oxide. Mechanochemical technologies is able to exclude these disadvantages.

For example, **V-Al-O** and **V-Ti-O** catalysts were prepared by mechanical activation of $V_2O_5$+$\gamma Al_2O_3$ and $V_2O_5$+$TiO_2$(anatase) mixtures [86]. According to $^{51}V$ NMR, the treatment results in disappearance of $V_2O_5$ signal and the appearance of a new signal with a chemical shift in the range –500 to -700 ppm whereas the interaction in nonactivated mixtures under heating at 510° is not complete and involves only a part of the vanadium atoms. The authors [87] noticed that under isothermal conditions between 565 and 615°C in nitrogen flowing atmosphere both the reduction of $V_2O_5$ and the transformation of anatase into rutile are topotactic reactions activated by a remarkable fit of the crystallographic patterns in contact at the $V_2O_5$-$TiO_2$ (anatase) interface.

In the study [88], XRD, $^{51}V$ NMR and EPR were used to show the principle possibility of mechanochemical synthesis of **V-Mo-O** compounds. The preliminary mechanical activation allows to prepare a number of V-Mo-O compounds excluding high temperature treatment. The authors suppose that they obtain new compounds with variable composition and common formula $V_2O_5 \cdot nMoO_3$, where n=2÷22. The compounds have crystal structure based on orthorombic $MoO_3$, in which Mo oxide layers are alternant with V oxide layers. It becomes possible due to steriochemical similarity of both oxides and the presence of weak bounded double polyhedral layers in 0k0 direction with V and Mo ions in the centre of polyhedron. Mechanical activation brings to the breakage of bonds in these structures in 0k0 direction between double polyhedron layers and, as a result, they are delaminated into packets with double polyhedron layers of Mo and V ions. Their broken end bonds are coordinationally unsatureated, therefore, the layers being mixed under mechanical activation tends to adjoint the neighbouring layers inspite of their cation composition. It brings to formation of V-Mo oxide structure consisting of alternant polyhedron layers with V and Mo ions (lamination mixing).

Bismuth vanadate, **$Bi_2VO_{5.5}$**, is a n=1 number of the Aurivillius family of oxides which represents a technologically important class of electronic materials. This compound is orthorhombic and ferroelectric at room temeperature and undergoes two marked phase transitions at 450 and 560°C. High temperature tetragonal (γ) phase of $Bi_2VO_{5.5}$ is a good ionic conductor. The use of $Bi_2VO_{5.5}$ for a wide variety of applications such as catalysts, gas sensors, posistors, solid state electrolytes, and positive electrode materials for lithium rechargeable batteries has already been demonstrated. However, alternate synthetic methods need to be developed to realize the full potential of this material.

The authors [89, 135] used mechanical activation of stoichiometric mixture of $Bi_2O_3$ and $V_2O_5$ to synthesize nanocrystalline bismuth vanadate powders. Structural evoluton of the desired $Bi_2VO_{5.5}$ through an intermediate phase ($BiVO_4$) was monitored by XRD, DTA and TEM investigations.

Iron vanadate, **$FeVO_4$**, is a prospective material for lithium rechargeable batteries and in catalysis. In [90] mechanical coactivation of iron and vanadium oxides was used to prepare intimate nanoscale mixture, similar to those prepared by soft chemistry. Reduction of this mixture at the same temperature and oxygen partial pressure conditions as of soft chemistry products (500°C and $10^{-25}$Pa) leads to formation of a nanometric vanadium ferrite with the only spinel phase. The characterization of the powders thus prepared was perfomed by X-ray diffraction, SEM, IR spectrometry, thermogravimetry and colourimetry. It was shown that the homogeneity of grain size and chemical composition is achieved if the initial oxides have similar grain size.

## 6.5. Manganates

**Lithium manganese spinel.** Lithium-manganese oxide compounds with spinel structure attract a great deal of interest in both theoretical and applicative aspects. A great ability of manganese ions to change easily their valence state in lithium manganese spinels and possibility for cation mixing are the origin of interesting physical properties of these compounds. On the other hand, lithium manganese oxide $LiMn_2O_4$ with spinel structure is a favourite among cathode materials with high energy capacity for rechargeable lithium batteries, since it is cheaper and less toxic. Structural characteristics, particle size and crystallinity of cathode materials play a significant role in their electrochemical behavior [91,92].

An important problem prohibiting $LiMn_2O_4$ from wider use as a cathode for lithium batteries is the unstable rechargeability, i.e. the amount of reversibly inserted lithium decreases gradually during cycling. It has been shown that the rechargeability could be greatly improved by using cathode materials with smaller particle size, thus providing the performance of the heterogeneous reaction of intercalation/deintercalation in the kinetic regime, and with disordered structure or nonstoichiometric composition. In spinels, disordering is achieved either through cation mixing (inverse spinel) or through cationic and anionic nonstoichiometry (deficient spinel).

At room temperature, $LiMn_2O_4$ has a cubic spinel-type structure, space group $Fd\bar{3}m$. The structure can be described as ideally consisting of a cubic close-packing arrangement of oxygen ions in *32e* sites. $Li^+$ ions occupy tetrahedral *8a* sites, while $Mn^{3+}$ and $Mn^{4+}$ ions occupy octahedral *16d* sites. Nearly at room temperature, $LiMn_2O_4$ undergoes a reversible transformation: cubic ($Fd\bar{3}m$) ⇔ orthorhombic (Fddd). The phase transition is accompanied by a change in the electrical conductivity, the latter being lower for the low-temperature phase.

Manganese ions in $LiMn_2O_4$ are present in two oxidation states. The increasing of $Mn^{3+}$ content during the intercalation (discharge) process brings to Jahn-Teller distortion. The asymmetric changing of cell parameters (the changing of c/a ratio for 10-16%) is followed by destruction of spinel structure. Therefore, last years a lot of researches have been directed toward the preparation of nonstoichiometric lithium-manganese spinels with more stable matrix. The doping by low valence cations (<4) decreases the amount of $Mn^{3+}$ and stabilizes cubic spinel structure. Monovalent Li doping seems to be the best since Li provides a larger valence difference relative to Mn than di- or trivalent cations. Hence, a small amount of dopant is effective. Nonstoichiometric spinels, enriched by Li, $Li_xMn_2O_4$ ($x>1$), were found to be more stable and to have great improvement in rechargeability on cycling [93].

Low-temperature methods provide yielding of lithium-manganese oxides with more oxidized manganese cations since manganese ions are stable preferentially as $Mn^{4+}$ at lower temperature.

In [94-98] mechanochemical method was used for the synthesis of high disperse stoichiometric and nonstoichiometric $Li_xMn_2O_4$ spinel starting from different manganese ($MnO_2$, $Mn_2O_3$, MnO) and lithium (LiOH, $LiOH{\cdot}H_2O$, $Li_2CO_3$) compounds.

Tthe reactions between LiOH and $Li_2CO_3$ with different manganese oxides can be viewed as acid-base interactions. According to Lewis classification, lithium compounds are bases whereas manganese oxides are acids of different strength. Acid-base properties of oxides significantly depend on the oxidation degree of cation. The acidity increases in the raw

$$MnO\ (2+) < Mn_2O_3\ (3+) < MnO_2\ (4+) < HMnO_4\ (7+).$$

**Mechanochemical synthesis of stoichiometric $LiMn_2O_4$ spinel.** According to X-ray study, the oxidation state of manganese ions greatly influenced on the kinetics of mechanochemical reaction of manganese oxides with LiOH and $Li_2CO_3$ (Fig. 6.22). $MnO_2$ reacts almost completely: reflections of $LiMn_2O_4$ only are present on X-ray patterns. All the patterns could be indexed using cubic spinel cell, space group $Fd\bar{3}m$. The broadening of reflections is connected with the decreasing of particle size and residual strain. On the other hand, no observable interaction between $Mn_2O_3$ and MnO with $Li_2CO_3$ occurs. The reflections of $Mn_2O_3$ are practically unchanged while the reflections of $Li_2CO_3$ are present at the background level. Under mechanical activation of the mixture of MnO with $Li_2CO_3$, a significant transformation of MnO into $Mn_3O_4$ occurs.

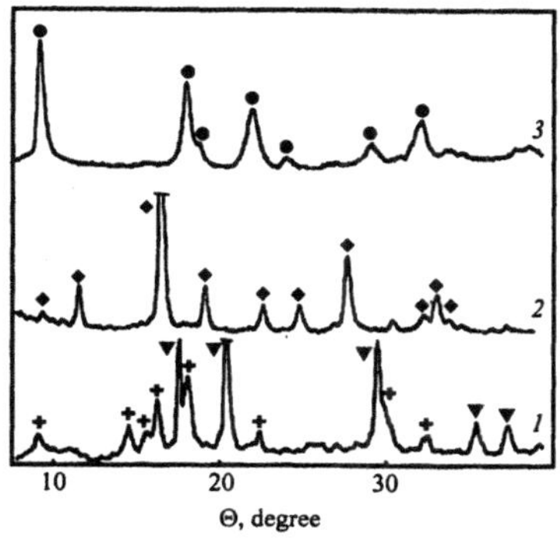


*Figure 6.22.* X-ray patterns of activated mixtures of $Li_2CO_3$ with different manganese oxides: 1 – MnO, 2 – $Mn_2O_3$, 3 – $MnO_2$. ● - $LiMn_2O_4$; ◆ - $Mn_2O_3$; ▼ - MnO; ✚ - $Mn_3O_4$.

According to thermal analysis, the activated samples have got some differences comparing with nonactivated ones. On DTA-curves of initial $MnO_2$+LiOH and $MnO_2+Li_2CO_3$ mixtures, endothermal effects corresponding to phase transitions of reagents ($T_{LiOH}$ = 440°C, $T_{Li2CO3}$ = =700°C, $T_{MnO2}$ = 540°C) are present (Fig. 6.23). These effects practically disappear after mechanical activation. New endoeffects are observed for activated samples at 50-600°C. Besides, a sharp exothermic peak appears on the heat curve of $MnO_2$+LiOH mixture activated for 10 min.Thus, considering X-ray and thermal analysis data, one can conclude that a short

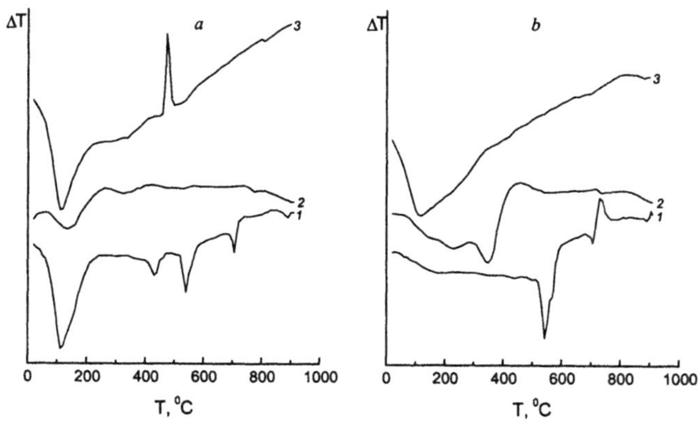


*Figure 6.23.* DTA heating curves of initial (1) and activated mixtures of $MnO_2$ with LiOH (a) and $Li_2CO_3$ (b): 2 – 1 min, 3 – 10 min.

mechanical activation of the mixtures (1 min in our case) brings to a fine grinding and mixing of reagents. As a result, a substantial acceleration of the interaction process between solids under the following heating at temperatures by 200-300°C lower than the melting point of lithium reagents takes place. Moreover, prolonged mechanical activation of mixtures brings to realization of direct solid state mechanochemical interaction of reagents, which is almost completed after 10 min.

The acceleration of solid state reactions under mechanical activation is promoted by the formation of the molecular-dense mechanocomposites. XPS and $^7$Li NMR spectroscopy revealed two different types of mechanocomposites initially formed in the activated mixtures of $MnO_2$ with LiOH and $Li_2CO_3$, depending upon structural and mechanical properties of lithium reagents.

$^7$Li NMR spectra of the mixtures of $MnO_2$ with LiOH and $Li_2CO_3$, activated for a short time in corundum jars, were found to be considerably different from each other [97]. In the first case, strong shifts to weaker magnetic fields are observed (Fig. 6.24). Such behavior is characteristic of a paramagnet, while lithium hydroxide is diamagnetic. Moreover, the value of the shift increases with the time of activation (see Table 6.1) and does not coincide with that in the final paramagnetic crystalline product, $LiMn_2O_4$ (470 ppm). The appearance of the strong paramagnetic shift in the $^7$Li NMR spectra of the activated mixtures of $MnO_2$ with LiOH was assigned to a spin density transfer on lithium atoms induced by the appearance of $Mn^{3+}$ ions and s-d band overlap. These processes, finally, bring about the formation of Li-O-Mn bonds.

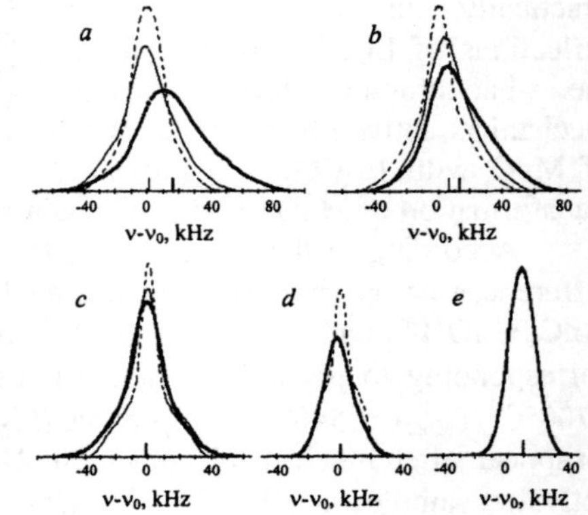


*Figure 6.24.* $^7$Li NMR spectra of the activated mixtures of $MnO_2$ with diamagnetic lithium compounds: a – LiOH; b - $LiOH{\cdot}H_2O$; c – $Li_2CO_3$; d - $Li_2SO_4{\cdot}H_2O$; e – a spectrum of $LiMn_2O_4$. Slim line – 1 min activation; thick line – 10 min activation; dashed line – a spectrum of initial lithium compound.

According to XPS [98], the Li/Mn ratio on the surface of particles of $MnO_2$+LiOH mixture activated for 1 min is twice as large as that after 10 min activation. But after 10 min of mechanical activation, the Li/Mn ratio is still 1.6 times larger than in Merck spinel (see Table 6.2), i.e. the surface layer of particles in activated mixtures is enriched by amorphous LiOH. On contrary, for 1 min activated mixture $MnO_2$+$Li_2CO_3$, Li/Mn ratio is equal to 0.8. The ratio decreases as the activation time increases.

*Table 6.1.* Values of nuclear quadrupolar constants (eQq/h), gravity centre schifts (δ) and half-widths (Δω) of lithium spectra of the initial and activated samples [97].

| Sample | eQq/h, kHz | δ, ppm | Δω, ppm |
|---|---|---|---|
| LiOH | 50 | 0±10 | 790±40 |
| $LiOH+MnO_2$,MA 1 min | - | 50±50 | 910±40 |
| $LiOH+MnO_2$,MA 10 min | - | 800±50 | 1500±40 |
| $Li_2CO_3$ | 63 | 0±10 | 560±40 |
| $Li_2CO_3+MnO_2$,MA 1 min | 90 | -60±50 | 560±40 |
| $Li_2CO_3+MnO_2$,MA 10 min | 90 | -80±50 | 830±40 |
| $LiMn_2O_4$ | 0 | 470±50 | 790±40 |

*Table 6.2.* Changing of surface layer content of the activated samples according to XPS data [98].

| Sample | O/Mn | Li/Mn |
|---|---|---|
| $LiOH+MnO_2$, 1 min MA | 3,0 | 3,2 |
| $LiOH+MnO_2$, 10 min MA | 2,5 | 1,6 |
| $LiOH+MnO_2$, 10 min MA and after heating at 800°C | 1,4 | 1,8 |
| $Li_2CO_3+MnO_2$, 1 min MA | 2,1 | 0,8 |
| $Li_2CO_3+MnO_2$, 10 min MA | 2,5 | 2,0 |
| $Li_2CO_3+MnO_2$, 10 min MA and after heating at 800°C | 1,3 | 1,7 |

Thus, one can propose that when LiOH, having a layered structure and exhibiting good plasticity, is used, a quick amorphization of this compound occurs, followed by its "smearing" on the surface of firmer $MnO_2$ particles. The adhesion forces at the contact of LiOH and $MnO_2$ particles exceed the cohesion forces inside $MnO_2$ particles. A thin film of amorphous LiOH acts as surfactant, accelerating the dispersion of $MnO_2$ particles. On the other hand, a number of high-reactive low-coordinated atoms on the surface of oxides is increased and facilitates the process of adhesion of hydroxide. This results in the formation of molecular-dense mechanocomposites. Dense contact between reagents brings to the formation of numerous micronuclei of a new phase of product. In the course of the following heating of both samples, the growth of nuclei of a new phase takes place. The more is the amount of micronuclei, the higher should be the dispersion of the final crystal product. In the case of $Li_2CO_3$, having ionic structure and exhibiting brittle properties, the brittle fracture of both components is observed. Fig. 6.25 shows the scheme of the formation of two different types of mechanocomposites, mentioned above.

*Figure 6.25.* The scheme of mechanocomposites formation in the activated mixtures of $MnO_2$ with LiOH (a) and $Li_2CO_3$ (b).

The revealed differences in the mechanism of the initial stages of the reaction affect the properties of final products, especially, the particle size and conductivity.

The value of specific surface area S of $LiMn_2O_4$ samples prepared by mechanical activation of LiOH+$MnO_2$ (Merck) mixture is equal to 90 $m^2/g$ and decreases to 12 $m^2/g$ after heating at 600°C. For samples prepared by mechanical activation of $Li_2CO_3$+$MnO_2$ mixture, S is equal to 32 $m^2/g$ and decreases to 8 $m^2/g$ after heating of the sample at 600°C.

According to Fig. 6.26, spinels, prepared by mechanical activation followed by heat treatment, are characterized by very low lattice parameter *a*, which increases as temperature rises. It is well known that the radius of $Mn^{4+}$ ions is lower than that of $Mn^{3+}$. $Mn^{4+}$ ions are more stable at lower temperatures. Thus, low values of *a* parameter for the samples with molar ratio Li/Mn=1/2 point to the increased amount of $Mn^{4+}$ ions comparing with that of stoichiometric spinel [93]. One can see that in the samples prepared from $Li_2CO_3$ by mechanical activation and heating at 450°C, the value of *a* and the amount of $Mn^{3+}$ ions are higher than in the case of LiOH samples because of reducing character of $CO_2$, eliminating in the course of interaction [98].

Besides, a low cell parameter of spinels as prepared could be caused by the presence of substantial disordering of cationic sublattices. One of well known lithium-manganese spinels with Li/Mn=1/2 molar ratio is $Li_2Mn_4O_9$, or $(Li_{0.89}\square_{0.11})_{tetr}[Mn_{1.78}\square_{0.22}]_{oct}O_4$ (where □ is a vacancy) in

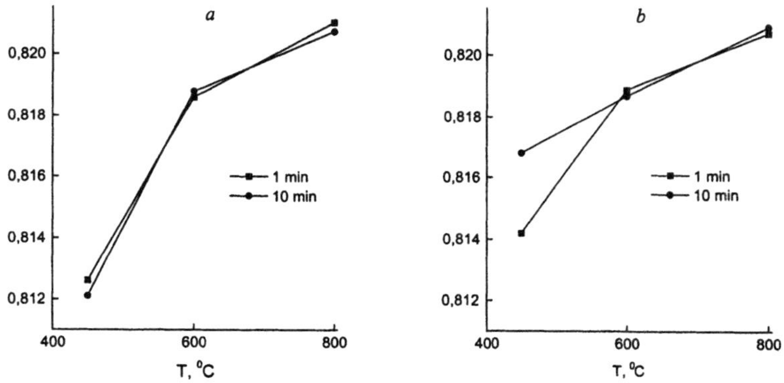


*Figure 6.26.* Lattice parameter of $LiMn_2O_4$ versus time of activation and temperature of heating: (a) prepared from $LiOH+MnO_2$; (b) prepared from $Li_2CO_3+MnO_2$.

spinel notation [93]. It is characterised by a pressed lattice (a=0.816 nm instead of 0.824 nm for $LiMn_2O_4$) and has all manganese ions in 4+ oxidation state. The composition of mechanochemically prepared spinels, probably, can be described by a similar formula $(Li_{1-x}\square_x)\cdot[(Mn^{4+}Mn^{3+})_{2-y}\square_y]O_4$ or $LiMn_2O_{4+z}$. On the increase of calcination temperature, the process of reduction of manganese ions and annealing of defects occur.

The excess of oxygen in mechanochemically prepared spinels is released under heating (Fig. 6.27). For all samples prepared at different temperatures this is a two-step process. The starting temperature of the second step coincides for all samples and is equal to 800°C. On contrary, the temperature of the first step is different and comprises 400°C for sample prepared at 450°C and 650°C – for the others. The cooling process is followed by mass gain. In the 800-1000°C region, the process is reversible and at lower temperatures is irreversible. According to [99], the value of irreversible mass loss can be associated with the changing of

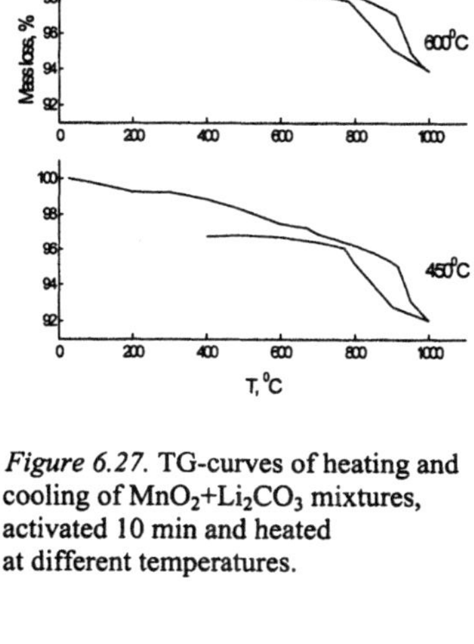


*Figure 6.27.* TG-curves of heating and cooling of $MnO_2+Li_2CO_3$ mixtures, activated 10 min and heated at different temperatures.

average oxidation degree of manganese ions, i.e. it must be equal to the amount of defects (cation vacancies). The estimations showed that the amount of vacancies for samples being heated at different temperatures varies from 0.07 to 0.02 per formula for tetra- (x) and from 0.14 to 0.04 for octa-positions (y) versus the temperature increasing. Note, that the samples heated even at 800°C conserve some amount of defects.

IR spectroscopy data point to the other type of disordering. Two asymmetric bands are present in the spectra of $MnO_2$+LiOH activated mixtures at 300-800 $cm^{-1}$ range (Fig. 6.28). They are degenerated for

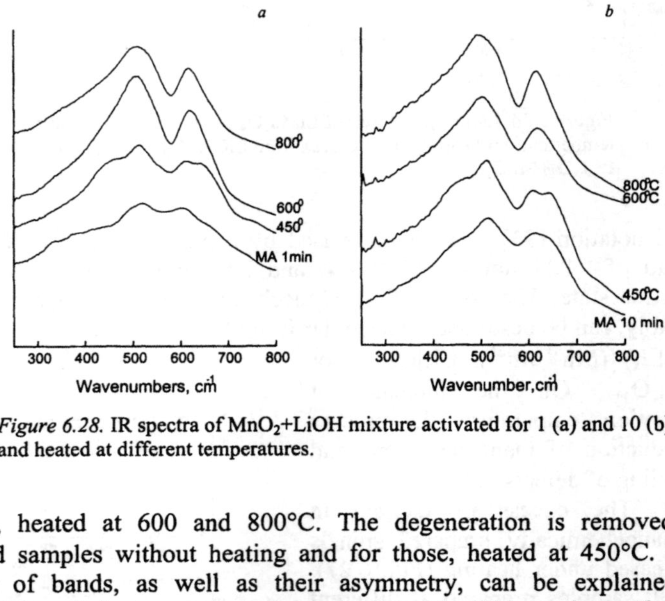


*Figure 6.28.* IR spectra of $MnO_2$+LiOH mixture activated for 1 (a) and 10 (b) min and heated at different temperatures.

samples, heated at 600 and 800°C. The degeneration is removed for activated samples without heating and for those, heated at 450°C. Such splitting of bands, as well as their asymmetry, can be explained by distribution of homogeneous manganese and lithium ions through nonequivalent crystallographic positions, i.e. the formation of mixed spinels.

The lithium ion distribution over different crystallographic sites was confirmed by $^7$Li NMR spectroscopy [96]. $^7$Li NMR spectrum of $LiMn_2O_4$, prepared by conventional solid state reaction, in which lithium ions occupy high symmetric tetrahedral 8a sites, is presented by a symmetric line with paramagnetic shift of 470 ppm toward weak field. The spectra of $LiMn_2O_4$ prepared by activation and heating of $MnO_2$+$Li_2CO_3$ mixture at 800°C, is presented by asymmetric line (Fig. 6.29).

To authors' opinion, the observed asymmetry arises from an increased number of neighboring oxygen atoms and indicates two different positions for lithium ions: tetrahedral and octahedral. It was considered that if all lithium ions occupy high symmetric tetrahedral 8a sites, NMR spectrum is presented by a narrow symmetric line. If all lithium ions occupy octahedral sites, the broadening of line should appear owing to electric field gradient. Estimations showed that ~5-7 % of lithium ions occupy octahedral sites.

**Mechanochemical synthesis of nonstoichiometric $Li_xMn_2O_4$ spinels.** The synthesis of nonstoichiometric spinels $Li_xMn_2O_4$ with variable composition was performed by mechanical activation of $xLi_2CO_3+4MnO_2$ mixtures with $0.21\leq x\leq 1.72$ [94]. XRD showed that the formation of lithium-manganese spinel occurs during mechanical activation in all mixtures, though the amount and the crystallinity of product are different. $LiMn_2O_4$ is the main crystalline product in the mixtures with $0.66\leq x\leq 1.21$. In the mixtures with $0.21\leq x<1$ $LiMn_2O_4$ coexists with $Mn_2O_3$ and in the mixtures with $1.45\leq x\leq 1.72$ – with $Li_2MnO_3$. After heating of activated samples at 600 and 800°C the crystallinity of spinel increases.

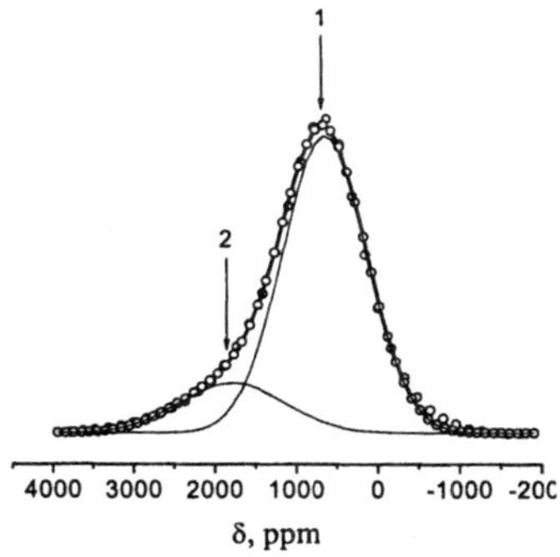


*Figure 6.29.* $^7$Li NMR spectrum of $LiMn_2O_4$ prepared from $MnO_2$+ $Li_2CO_3$ mixture activated 10 min and heated at 800°C: 1 - Li ions in tetrahedrons; 2 - in octahedrons.

Fig. 6.30 shows the dependence of lattice parameter *a* versus *x* in samples as prepared. In the samples with $0.21\leq x\leq 1.00$ the lattice parameter is practically constant and has a value of 0.822 nm (600°C) and 0.824 nm (800°C) and sharp decreases to 0.814 nm (600°C) and 0.818 nm (800°C) at x>1. The decreasing of lattice parameter for the samples with lithium excess can be connected with the increasing of average oxidation state of manganese. Note, that no tetragonal distortion of the spinel lattice with increasing of x is observed on X-ray patterns. It points to stabilization of spinel structure due to the increasing of Mn (4+) content. The spinels as prepared were investigated by complex impedance technique. Conductivity of unsintered pellets is relatively low, $10^5$-$10^{-6}$ S/cm, and strongly depends on starting reagents and the molding pressure P. Pellets of $LiMn_2O_4$ samples obtained from $Li_2CO_3$ have higher conductivity

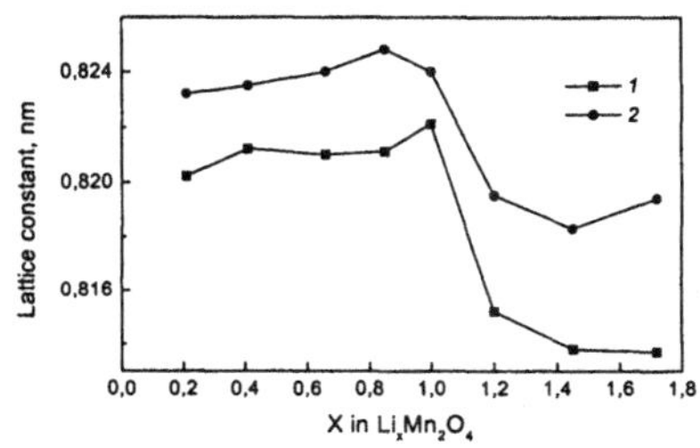


*Figure 6.30.* Effect of Li content in activated $MnO_2$+$Li_2CO_3$ mixtures on the lattice constant of $LiMn_2O_4$ prepared after heating at 600 (1) and 800°C (2).

than those prepared from LiOH. Conductivity monotonically raises with the increasing of molding pressure from 300 to 900 MPa. Heat treatment leads to the rise of conductivity. Pellets sintered at T > 600°C have conductivity of about $1 \cdot 10^{-4}$ S/cm irrespective of starting material.

Thus, the method of mechanical activation, combined with short heating at moderate temperatures, can be used for preparing of lithium manganese spinels with unusual properties. Mechanochemically prepared spinels are characterized by high dispersion, poor crystallinity, strongly compressed lattice, increased amount of manganese ions in 4+ oxidation state and, at least, two types of disordering: cation vacancies and cation mixing. The dispersion, the amount of defects and $Mn^{4+}$ ions decrease and crystallinity increases under following thermal treatment. The nature of reagents and the conditions of mechanical activation influence the parameters of spinel, especially at moderate temperatures. This provides possibility to regulate the properties of final products, providing 3V or 4V cathode materials. To our opinion, high dispersed and disordered $LiMn_2O_4$ as prepared should display good cathode characteristic at high rate discharged cells.

**Lanthanum manganese perovskite.** $LaMnO_3$ and related compounds are widely studied due to their interesting physical properties. The widespread interest in this system was generated, first, by the emergence of strontium-doped lanthanum manganite as the cathode material for the solid oxide fuel cell and, more recently, by the discovery of colossal magnetoresistance in the compounds with mixed Mn valence. At last, these compounds were found to be promising high-temperature monolithic catalysts of such important reactions as fuel combustion, oxidation of ammonia to give nitrogen oxides, chlorinated hydrocarbons incineration, etc.

The authors [100] investigated mechanochemical synthesis of these perovskites using transition oxides in different oxidation state and lanthanum oxide or carbonate as initial compounds. The activated samples were then heated at temperatures above 500°C. The degree of interaction was shown to depend upon the oxidation degree of transition metal ions in oxides. The higher oxidation degree, i.e. the acidity of oxide, is preferable.

Specific catalytic activity in the reaction of CO and butane oxidation was found to be a function of the calcination temperature with its maximum at 800°C. SIMS data demonstrate that the middle- and high-temperature samples differ by the transition metal surface concentration. In some cases, the high level activity of middle-temperature disordered samples can be assigned to segregation of $MnO_x$ on the surface of the particles.

## 6.6. Ferrites

**Lithium ferrites.** Lithium ferrite $LiFe_5O_8$ is widely used in microwave devices and is considerably superior to garnets with respect to stress sensitivity and cost. The preparation of $LiFe_5O_8$ by conventional ceramic methods employing high temperatures often produces low quality materials. The volatility of lithium and oxygen above 1000°C severely effects the magnetic properties and resistivity. The lost of $Li_2O$ results in the formation of $\gamma$-$Fe_2O_3$, which precipitates out as $\alpha$-$Fe_2O_3$ under cooling, and leads to a lower magnetization. If oxygen is lost in excess of $Li_2O$, some ferric ions are reduced to ferrous. This produces a solid solution of $Fe_3O_4$ in lithium ferrite and the presence of $Fe^{2+}$ affects the anisotropy and magnetization in the same way as would a paramagnetic substituent. The presence of $Fe^{2+}$ and $Fe^{3+}$ ions on octahedral sites also brings to a valence exchange process, which reduces the resistivity by as much as $10^4$ and increases microwave losses.

The problems, which arise from the volatilization of lithium and oxygen, are possible to overcome, by employing low-temperature methods. These methods could produce homogeneous monophase fine particles of materials which sinter to near theoretical densities at relatively low temperature.

In [101] the synthesis and alteration of $\alpha$-$LiFeO_2$ by mechanochemical activation were studied. A mixture of $\gamma$-FeOOH and $LiOH{\cdot}H_2O$ was ground for different time intervals in a planetary ball mill (Retsch) equipped with a 250 ml agate jar and five balls of the same material, 1 cm in diameter.

The thermal evolution of nonground mixture is correspondent to the following transformations:

$$LiOH{\cdot}H_2O \rightarrow LiOH + H_2O \ (140°C)$$
$$2\ \gamma\text{-}FeOOH \rightarrow \gamma\text{-}Fe_2O_3 + H_2O \ (220°C)$$
$$2\ LiOH + \gamma\text{-}Fe_2O_3 \rightarrow 2\ \alpha\text{-}LiFeO_2 + H_2O \ (430°C).$$

For the sample ground for 7 h, endothermic peak at 430°C associated with the formation of $LiFeO_2$ disappears. Moreover, the magnitude of the remaining endothermic peaks and the weight loss decrease markedly. This indicates that Li and Fe compounds undergo a progressive mechanochemical dehydration in which the increase in the mobility of $Li^+$ ions, caused by the shearing stress and created under the impact of the balls inside the jar, plays a definite role.

XRD shows the broadening of line profiles at the initial stages of mechanical activation and the occurrence of crystalline modifications of $LiFeO_2$ after 7 h grinding. A more prolong mechanical activation improves $LiFeO_2$ crystallinity and causes the mechanochemical phase transition $\beta$-

$LiFeO_2 \rightarrow \alpha\text{-}LiFeO_2$. The partial transformation of $\alpha\text{-}LiFeO_2$ into the spinel phase $\alpha\text{-}LiFe_5O_8$ was also observed as well as in the ground $\alpha\text{-}LiFeO_2$ sample prepared by firing of a stoichiometric mixture of $Li_2CO_3$ and FeOOH in air at 850°C for 48 h. It was found that atomic ratio Li/Fe becomes smaller than unity as the grinding time increases. The increase in the $Li^+$ ion mobility within the solid under mechanical activation brings to the following reactions:

$$LiFeO_2 \rightarrow Li_{1-x}FeO_{2-x/2} + x/2\ Li_2O$$
$$Li_{1-x}FeO_{2-x/2} \rightarrow x/4\ LiFe_5O_8 + (4-5x)/4\ LiFeO_2.$$

The water molecules produced during mechanochemical dehydration of lepidocrocite facilitate the lithium mobility in a way similar to that found when LiOH melts and reacts with $Fe_2O_3$. The structural transformation is explained within a model in which $Li^+$ vacancies created during grinding promote the migration of $Fe^{3+}$ ions from octahedral to tetrahedral sites.

The authors [102] investigated the influence of initial reagents on mechanochemical synthesis of $LiFeO_2$. They found that $LiFeO_2$ is formed as a result of mechanical activation of mixtures of $LiOH{\cdot}H_2O$ with $\alpha$-FeOOH, $\gamma$-FeOOH, $\gamma\text{-}Fe_2O_3$ and $\alpha\text{-}Fe_2O_3$ in high energetic planetary mill. No noticeable difference between reactivity of Fe-containing compounds was observed. The product is formed when either LiOH or lithium acetate are used instead of $LiOH{\cdot}H_2O$, but the rate of the reaction being less. On contrary, lithium compounds with ionic lattice, such as $Li_2CO_3$, $Li_2SO_4$, $Li_2SO_4{\cdot}H_2O$, do not react mechanochemically to yield $LiFeO_2$. To authors' opinion, under shearing stress in activators, lithium compounds with layered structure are fragmentated along the layers that brings to easy "layer by layer" mixing of reagents as a result of plastic deformations. In this case the role of $Li^+$ diffusion in the reaction of solid state synthesis becomes inessential, that allows to perform the process at room temperature. When reagents with strong covalent or ionic lattice without planes of "facile plastic defirmation" are used, mechanical stresses bring to brittle fracture and fine mixing only. This accelerates the kinetics of the reaction at the following heat treatment. The role of water molecules in the lattice of compounds, which yield $LiFeO_2$ under mechanical activation, consists, probably, in sponginess of their structure and decrease of their mechanical strength at atom-molecular level. It was stated that no crystallization of $LiFe_5O_8$ directly in mill' jars occurs whenever pairs of reagents being used.

The authors [103] studied the influence of reagent biography and reaction conditions on kinetics of $LiFeO_2$ and $LiFe_5O_8$ synthesis by applying mechanical activation.

Under thermal conditions $LiFe_5O_8$ formation proceeds into 2 steps:

$Li_2CO_3 + Fe_2O_3 \rightarrow 2\ LiFeO_2 + CO_2$
$LiFeO_2 + 2Fe_2O_3 \rightarrow LiFe_5O_8$.

The preliminary mechanical activation of $\alpha$-$Fe_2O_3$ and $\gamma$-$Fe_2O_3$ significantly increases the kinetics of these reactions. Comparing the different activity of $\alpha$- and $\gamma$-$Fe_2O_3$, the authors proposed that the elevated activity of $\gamma$-$Fe_2O_3$ in both reactions is associated not with its higher specific surface area, but with the close conformity of oxygen sublattices of $\gamma$-$Fe_2O_3$ and $LiFe_5O_8$ requiring only the counter diffusion of Li and Fe cations for the reaction to take place. Additionally, the possibility of more fast diffusion of the iron cations in $\gamma$-$Fe_2O_3$ due to the higher number of vacant octahedral sites facilitating this process, as compared with $\alpha$-$Fe_2O_3$, should be also taken into account.

**Magneisum ferrite spinel.** $MgFe_2O_4$ is one a representative of soft ferrites extensively used for high-frequency applications. Fully dense bulk ferrites have been synthesized by low-temperature sintering of fine powders, by thermal decomposition and by co-precipitation. Preparation from ultra-fine powder is more advantageous since the composition can be more easily controlled, and the electrical and thermal properties are improved as a result of the reduced grain size.

The authors [104] synthesized $MgFe_2O_4$ by mechanical grinding of stoichiometric amounts of crystalline $Fe_2O_3$ and amorphous $(MgCO_3)_4{\cdot}Mg(OH)_2{\cdot}5H_2O$ in high energetic centrifugal and vibration mills. The reaction was studied for two conditions: without solvent (dry milling) and with toluene (wet milling). It was shown that the reaction performed without solvent is faster than using toluene.

**Calcium ferrite perovskite.** Perovskite compounds, intensively studied recent years, possess unique physical and chemical properties, since their oxygen stoichiometry can be varied over a wide range by various treatments.

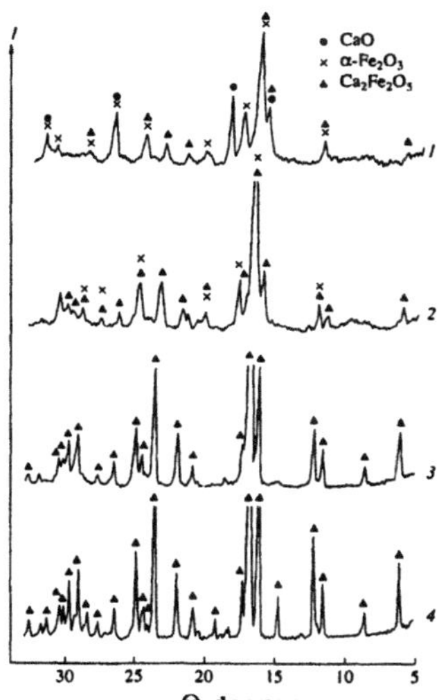

*Figure 6.31.* X-ray patterns of the $2CaO$+$\gamma$-$Fe_2O_3$ mixture after activation for 2.5 (1) and 10 (2) min and heated at 600 (3) and 1000°C (4).

$Ca_2Fe_2O_5$, a perovskite ferrite, is widely used as a catalyst for oxidation of carbon monoxide, hydrocarbons, ammonia, etc. Conventional high-temperature synthesis does not provide catalysts with high specific surfaces and, hence, with high catalytic activity. The lowing of temperature preparation can be achieved by preliminary mechanical activation of reactant mixture. It was reported [105] that thermal synthesis of $Ca_2Fe_2O_5$ from mechanically activated mixture of CaO and $\alpha$-$Fe_2O_3$ is markedly accelerated, especially at moderate temperatures.

The authors [106] investigated mechanochemical synthesis of dicalcium ferrite with perovskite structure from anhydrous or hydrated calcium and iron oxides in high energetic planetary mill with stainless jars and balls by XRD, DTA, IR and Mössbauer spectroscopy. According to X-ray, crystalline $Ca_2Fe_2O_5$ was formed directly from the mixture of CaO and $\alpha$- or $\gamma$-$Fe_2O_3$ within 2-3 min of mechanical activation (Fig. 6.31). In the latter case $\gamma$-$Fe_2O_3$ first transformed into $\alpha$-$Fe_2O_3$. In the IR spectra of activated mixtures, the absorbance bands of $Ca_2Fe_2O_5$ appear.

Figure 6.32 shows the Mössbauer spectrum, supplied by its decomposition, of $2CaO+\alpha$-$Fe_2O_3$ mixture activated for 2.5 min. One can see that the spectrum is a superposition of two sextets and two doublets. The sextets are due to $Fe^{3+}$ ions in octahedral and tetrahedral coordination, respectively. Two doublets are attributed to $Fe^{3+}$ ions in amorphous phases. Thus, according to Mössbauer spectroscopy data, the formation of crystalline $Ca_2Fe_2O_5$ occurs simultaneously with the formation of amorphous phase. The amount of $\alpha$-$Fe_2O_3$ decreases with activation time, and after 5 min of mechanical activation $\alpha$-$Fe_2O_3$ is not observed. The amount of the amorphous phase increases throughout the process, while the amount of crystalline $Ca_2Fe_2O_5$, calculated from Mössbauer peak areas, goes through the maximum (Fig. 6.33).

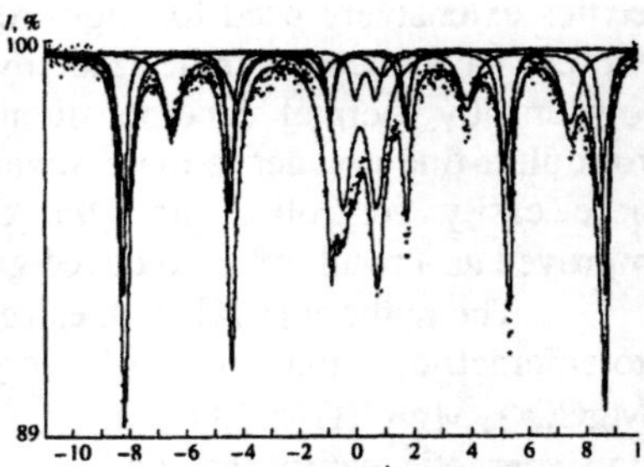


*Figure 6.32.* Mössbauer spectrum and its decomposition for $2CaO+\alpha$-$Fe_2O_3$ mixture activated 2.5 min (dots: measured spectrum).

After calcination of the activated samples at 400-1100°C for 2-4 h, only monophase crystalline calcium ferrite phase was observed. $S_{sp}=1\div3$ $m^2/g$. Note, that without mechanical activation anhydrous oxides do not react completely even at 1000°C, giving a mixture of $Ca_2Fe_2O_5$ and $CaFe_2O_4$, while monophase $Ca_2Fe_2O_5$ is formed only after calcination at 1100°C for 100 h.

On contrary, no reaction occurred during mechanical activation of the mixture of hydroxides, $\alpha$-FeOOH (goethite) and $Ca(OH)_2$. Only amorphization of reagents and partial decomposition of $\alpha$-FeOOH to $\alpha$-

$Fe_2O_3$ were observed. The amount of crystalline $\alpha$-$Fe_2O_3$ in the 10 min activated mixture is equal to 5-10 %. Thermal analysis shows that dehydration temperature of $Ca(OH)_2$ decreases from 520 to 440°C in activated mixtures. Nevertheless, the fine product, $Ca_2Fe_2O_5$, is formed after heat treatment at 600°C.

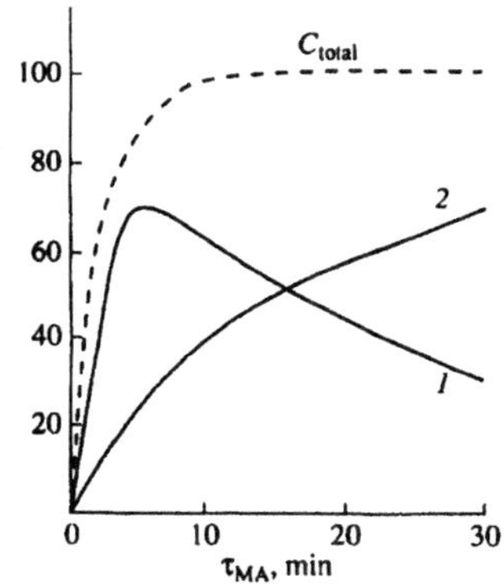

*Figure 6.33.* Amounts of crystalline (1) and amorphous (2) $Ca_2Fe_2O_5$ versus time of activation for $2CaO$+$\alpha$-$Fe_2O_3$ mixture.

For the $Ca(OH)_2$ + $\alpha$-FeOOH mixture activated for 10 min, $S_{sp}$ = 3.7 $m^2/g$, and the size of coherently scattering regions is of a few tens of nanometers. This suggests that dense aggregates of small particles are likely to be formed. The formation of dense aggregates, in which initial particles are in dense contact, considerably facilitates subsequent thermally induced interaction. At 600°C, these aggregates disintegrate, causing an increase of $S_{sp}$ up to 13 $m^2/g$. At higher temperature, sintering of small particles is observed leading to the decrease of specific surface area.

Crystalline $Ca_2Fe_2O_5$ obtained by thermal synthesis, appeared to be unstable to mechanical treatment. According to Mössbauer spectroscopy data, the amount of the amorphous phase grows rapidly to 20-25 % during mechanical activation (75-80 % of crystaline phase) (Fig. 6.33). The rate of amorphization process increases if mechanical activation of $Ca_2Fe_2O_5$ is carried out in the presence of small amounts of doped water (Fig. 6.34). Thus, the final result of mechanical activation of anhydrous oxides mixtures is determined by the correlation between the processes of formation of crystalline $Ca_2Fe_2O_5$ and its amorphization.

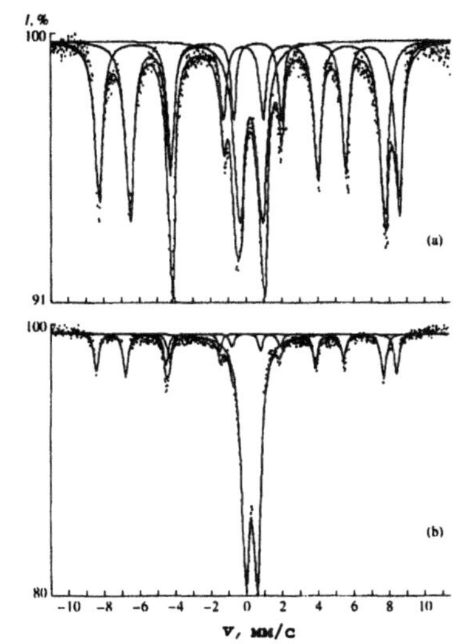

*Figure 6.34.* Mössbauer spectra and their decomposition for $Ca_2Fe_2O_5$ activated for 10 min at the absence (a) or presence (b) of water (dots: measured spectra).

The solubility of samples as prepared depends on calcination temperature. After calcination at low temperature (400-500°C), samples are dissolved in solutions with a low acidity even at room temperature. After calcination at high temperatures (900-1000°C), they can be dissolved only in stronger acid media at 80-90°C. Hence, the product formed at low calcination temperatures is defective; this influences on its ability to dissolution as well as on catalytic activity.

**Strontium ferrite perovskite.** The nonstoichiometric strontium ion oxide $SrFeO_x$ ($2.5 \leq x \leq 3.0$) is a well known mixed oxygen conductor. Being a solid state electrolyte itself, doped with rare earths or transition metals, it forms materials, which exhibit even greater ionic conductivity. These oxides are used as materials for self-supported oxygen conductivity membranes.

$SrFeO_x$ is usually produced via solid state reaction between strontium and iron compounds. This method requires prolong firing at high temperatures (1200°C for about 24 h). This is followed by subsequent grinding and calcination.

The influence of high energetic ball milling of reactants on a heat activated synthesis of $SrFeO_{2.5}$ from hematite and strontium carbonate was investigated in [107]. The authors studied dry and wet milling (5 ml of deionized water per 10 g of the mixture) considering a wet milling to be the faster way to reduce the size of reactants' grains and, therefore, to bring the synthesis temperature down. They stated that while heating of nonactivated mixture to 1000°C, the products consist of a mixture of orthorombic brown millerite $SrFeO_{2.5}$, cubic strontium oxide SrO and $Sr_7Fe_{10}O_{22}$. For milled samples, the reaction yielded only $SrFeO_{2.5}$ and there was no visible iron contamination from the jars and balls neither in form of metallic Fe nor iron oxides. The reactivity at lower temperatures increased with milling time and was higher for wet processed powders. For nonactivated samples the reaction was completed only after 35 h of heating at 1200°C.

The authors noticed also that the presence of $Fe_2O_3$ in the mixture with $SrCO_3$ lowers the decomposition temperature of the latter. The temperature decreases with increasing hematite to carbonate ratio in the mixture. This means that the temperature strongly depends on a contact area of $SrCO_3$ with $Fe_2O_3$. Milling causes partial transformation of hematite to magnetite, which results in a smaller mass loss of the samples during calcination.

Thus, mechanical treatment of the reactant mixture does not trigger the reaction. However, it significantly lowers the temperature required for the reaction by increasing the contact area between reactants. It also produces very uniform mixtures that requires much shorter heat treatment to attain composition equilibrium.

**Barium hexaferrite**. Barium hexaferrite, $BaFe_{12}O_{19}$, is used as a ceramic permanent magnet. It is conventionally prepared by firing an appropriate mixture of $\alpha$-$Fe_2O_3$ with $BaCO_3$ at high temperatures (1150-1250°C). To reduce the particle size, thus obtained ferrite should be ground.

The coercive force of these magnetic materials is much less than predicted by theory. Different methods to prepare better material have been proposed.

The authors [108] investigated mechanochemical synthesis of $BaFe_{12}O_{19}$ starting from stoichiometric mixture of lepidocrocite ($\gamma$-FeOOH) and $Ba(OH)_2 \cdot 0.3H_2O$. Besides $BaFe_{12}O_{19}$, samples doped by $Co^{2+}$ and $Ti^{4+}$, according to formula $BaCo_xTi_xFe_{12-2x}O_{19}$, where $0.5 \leq x \leq 0.75$, were also synthesized.

It was shown that the intermediate X-ray amorphous products, possessing water in their structure, were formed under interaction of initial reagents. Authors proposed that the amorphous state was stabilized by water and $CO_2$ molecules. Therefore, the final crystal product, $BaFe_{12}O_{19}$, was formed only after full elimination of water from the sample. According to electron microscopy data, the product as prepared has two kinds of particles, one of them being of 0.1-2 μm in size and the other ones – of 10-50 μm. The latter, probably, represent the aggregates of articles. According to magnetic measurements, thus prepared materials could be promising in practice.

## 6.7. Cobaltites

**Lithium cobaltite $LiCoO_2$.** $LiCoO_2$ is used as a cathode material in advanced rechargeable lithium batteries [109]. It is well known that $LiCoO_2$ can be prepared in 2 modifications, i.e. a high-temperature (HT) and a low-temperature (LT). The structure of HT-$LiCoO_2$ is related to trigonal space group $R\bar{3}m$, with oxygen anion occupying *6c* sites and Li and Co cations occupying *3a* and *3b* octahedral sites, respectively, and with $c/a$=4.99 [110]. Li layers alternate with Co layers between layers of oxygen. LT-$LiCoO_2$ is believed to adopt a spinel-like structure, consisting of a cubic-close packed oxygen network with alternating cation layer of composition 0.75 Co, 0.25 Li and 0.75 Li, 0.25 Co perpendicular to each of the four cubic (111) directions [111]. The space group is $Fd\bar{3}m$; Li and Co ions are in *16c* and *16d* sites, respectively, and O ions are in *32e* sites; $c/a$=4.90. It was shown that HT-$LiCoO_2$ has a larger capacity and less cycling fading [112].

Electrochemical behavior of $LiCoO_2$ strongly depends on the method of synthesis. Last years a great deal of attention was devoted to preparation of metastable lithium-transition metal oxides with unusual ordering and valence states that cannot be obtained by conventional high-temperature technique. The crystal and electronic structure of such products, as expected, must be more stable to intercalation/deintercalation of lithium ions in the course of discharge/charge processes, thus, providing better electrochemical behavior and preventing structural distortion. Li ions mobility is mainly related to the geometry of diffusion pathways and maximum Li uptake is a function of ionic sites that can accommodate Li ions and/or the number of electronic sites available for the corresponding electrons.

The particle size of the cathode material plays a very important role in their electrochemical behavior [113]. Smaller particles provide the performance of the heterogeneous reaction of intercalation/deintercalation in the kinetic regime. One of the problem is to enhance electronic conductivity of cathode materials. In this case, electronic additives (e.g., carbon) could be excluded.

When using low-temperature methods, it was stated that cobalt hydroxides ($Co(OH)_2$ or CoOOH) are usually more reactive, comparing with $Co_3O_4$. Higher reactivity of these compounds in the synthesis of layered $LiCoO_2$ is based on similarity of their structure ($R\bar{3}m$ for CoOOH and $P\bar{3}m1$ for $Co(OH)_2$). Under interaction in the mixtures with LiOH, they exhibit their amphotheric properties. In [114] HT-$LiCoO_2$ was prepared by freeze-drying of finely divided cobalt hydroxide – lithium hydroxide mixtures, obtained by precipitation. The authors [115] prepared HT-$LiCoO_2$ powders at temperatures as low as 100°C through the use of a two days ion exchange reaction between CoOOH and an excess of $LiOH{\cdot}H_2O$ in the presence of small amount of added water as well as in autoclave. The product possessed some amount of carbonate and hydroxyl species.

$LiCoO_2$ can be prepared by mechanical activation. It was found [116] that a phase, structurally related to HT-$LiCoO_2$, is obtained by grinding of stoichiometric mixtures of $LiOH{\cdot}H_2O$ and $Co(OH)_2$ for 10 h. This process takes place without the occurrence of $Co_3O_4$ as an intermediate, in contrast to the interaction in nonactivated mixture under heating. More prolonged mechanical treatment resulted in the elimination of lithium from $LiCoO_2$ structure. $Co_3O_4$ was observed in the activated sample after following heat treatment. This points out to the presence of poorly crystallized $Co_3O_4$ and/or a small amount of $H^+$ replacing $Li^+$ in the structure of $LiCoO_2$ after grinding for 40 h, and the formation of $Li_{1-x}Co^{3+}{}_{1-y}Co^{2+}{}_{y}O_{2-x/2-y/2}{\cdot}nH_2O$.

The authors [116] showed that the structure of high thermal stability, such as the ordered layered structure of HT-$LiCoO_2$, can be easily distorted with the help of the shearing stress induced by grinding, according to the following reaction:

$$6LiCoO_2 \rightarrow 2Co_3O_4 + 3Li_2O + 0.5O_2.$$

Besides, $Li_2O$ may be mainly deposited as a layer on the surface of the balls and the container.

The instability of HT-$LiCoO_2$ to mechanical activation was also observed in [117].

On contrary, You et al. [118] showed that LT-$LiCoO_2$ can be easily prepared by mechanical activation of the mixture of anhydrous $Li_2O_2$ and CoO, but the amount of contaminations was significantly higher due to mechanical properties (hardness) of initial reagents.

In [119] the mechanochemical interaction of LiOH with $Co(OH)_2$ and CoOOH, followed by thermal treatment at different temperatures, to yield $LiCoO_2$ was investigated by means of XRD, IRS and DRS. Although both Co compounds have a structure similar to the structure of $LiCoO_2$, the oxidation degree of Co ions is different. Thus, the reaction of $LiCoO_2$ formation using $Co(OH)_2$, contrary to CoOOH, can proceed only in the presence of free oxygen, i.e. via redox stage:

$$4\ LiOH + 4\ Co(OH)_2 + O_2 \rightarrow 4\ LiCoO_2 + H_2O$$
$$LiOH + CoOOH \rightarrow LiCoO_2 + H_2O$$

Besides, according to the statements, developed in Chapter 2, the acidity of CoOOH should be higher than of $Co(OH)_2$ because of higher oxidation degree of Co ions.

Mechanical activation was carried out in high energetic planetary activator in air. The time of mechanical activation varied from 1 to 10 min. The activated powders were then heated at 400, 600 and 800°C in air. On X-ray patterns of activated mixtures one can observe only broadened reflections of Co compounds (Fig. 6.35); the reflections of LiOH are absent,

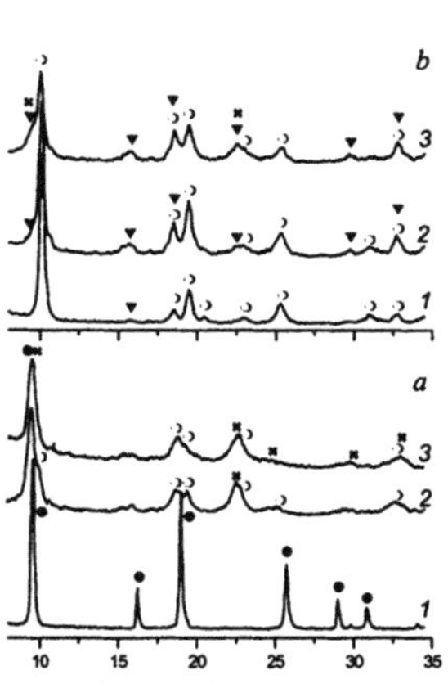


*Figure 6.35.* X-ray patterns of LiOH+ $Co(OH)_2$ (a) and LiOH+CoOOH (b) mixtures activated 1 (2) and 10 (3) min compared to initial $Co(OH)_2$ and CoOOH (1).
● - $Co(OH)_2$, ○ - CoOOH, ▼ - $Co_3O_4$, ✖ - $LiCoO_2$.

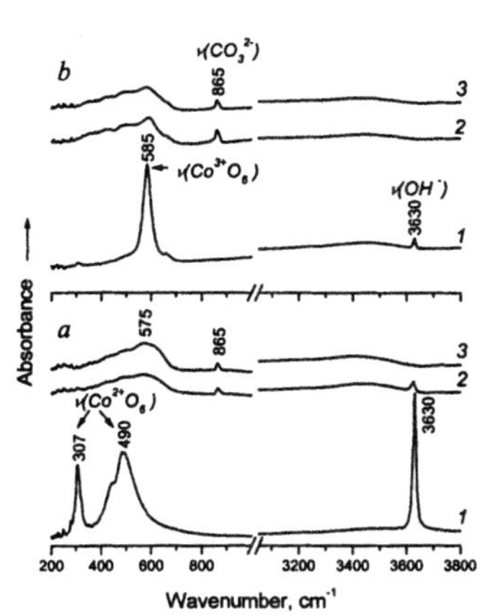


*Figure 6.36.* IR spectra of LiOH+ $Co(OH)_2$ (a) and LiOH+CoOOH (b) mixtures activated 1 (2) and 10 (3) min compared to initial $Co(OH)_2$ and CoOOH (1).

probably, due to its amorphization and weak scattering ability of Li ions. In the mixture with $Co(OH)_2$, a partial oxidation of $Co(OH)_2$ with the formation of CoOOH occurs. On contrary, the traces of $Co_3O_4$ appear on X-ray pattern of activated mixture with CoOOH, pointing to its partial reduction. Some weak reflections on X-ray patterns of both mixtures can be assigned to $LiCoO_2$.

In IR spectra of both activated mixtures (Fig. 6.36) one can observe new bands at 500-700 $cm^{-1}$ range corresponding to stretching vibrations of $Co^{3+}O_6$ bonds, probably, of nuclei of product - $LiCoO_2$.

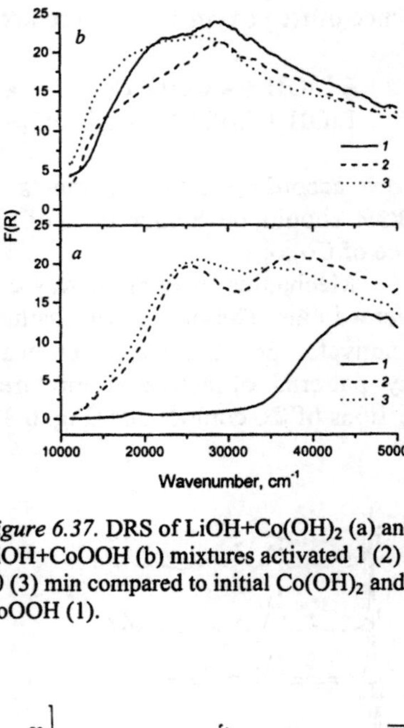


*Figure 6.37.* DRS of LiOH+$Co(OH)_2$ (a) and LiOH+CoOOH (b) mixtures activated 1 (2) and 10 (3) min compared to initial $Co(OH)_2$ and CoOOH (1).

Heat treatment of activated samples brings to completion of chemical interaction and the formation of homogeneous product - $LiCoO_2$. The width of reflections decreases and the intensity increases versus temperature as a result of crystallization and particle growth. The presence of splitting of reflections 006 and 012; 018 and 110, as well as *c/a* ratio equal to 4.99, point to the formation of HT-modification of $LiCoO_2$ even at 400°C. Note, that heating of nonactivated mixture at 400°C brings to the formation of a mixture of HT- and LT-$LiCoO_2$.

More detail information on the change of electronic state of cobalt ions in activated samples and final products prepared at different temperatures was obtained by diffusion reflectance electron spectroscopy (DRS). In DRS spectra of LiOH+$Co(OH)_2$ mixture (Fig. 6.37*a*), the intensity of background increases above 15000 $cm^{-1}$ as a result of strong distortion of $Co(OH)_2$ structure. New bands at 25000 $cm^{-1}$, corresponding to

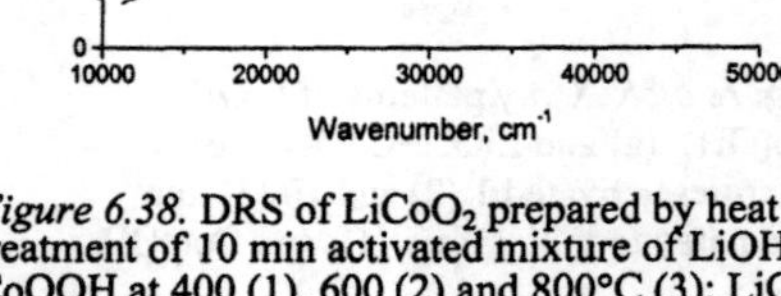


*Figure 6.38.* DRS of $LiCoO_2$ prepared by heat treatment of 10 min activated mixture of LiOH+ CoOOH at 400 (1), 600 (2) and 800°C (3); $LiCoO_2$ commerc. (4).

$[Co^{3+}]_{Oh}$, appear due to partial oxidation of $Co^{2+}$ ions. In spectra of LiOH+CoOOH activated mixture (Fig. 6.37*b*), the band at 27000-28000 $cm^{-1}$, corresponding to low-spin $[Co^{3+}]_{Oh}$, is the most intensive, while the band at 22000 $cm^{-1}$, present in the spectrum of initial CoOOH and assigned to $Co^{3+}$ ions in more distorted octahedrals (surrounded by OH groups), becomes weaker.

Under heating, the spectra are noticeably changed. New, more clear bands of localized centers (ions) appear. In DRS spectra of $LiCoO_2$ samples, prepared by mechanical activation of LiOH+CoOOH mixture for 10 min and subsequent heat treatment at 400 and 600°C (Fig. 6.38), one can see absorption bands at 28000, 18000-22000 and 15000-17000 $cm^{-1}$, which correspond to $[Co^{3+}]_{Oh}$, $[Co^{2+}]_{Oh}$ and $[Co^{2+}]_{Td}$, respectively. After heating at 800°C, the band at 22000 $cm^{-1}$ practically disappears, while the band at 15000-17000 $cm^{-1}$ becomes more intensive, the latter pointing to the presence of $Co_3O_4$ with spinel structure as an admixture. Note, that in the samples as prepared, the main band, corresponding to $[Co^{3+}]_{Oh}$, is present at lower frequency range comparing with commercial $LiCoO_2$, most evidently, as a result of formation of less perfect octahedrons $Co^{3+}O_6$, connected with the weakening of crystal field force.

**Lanthanum cobaltite $LaCoO_3$.** $LaCoO_3$ is high-active catalyst for oxidation of CO. It is also used as an electrode in high-temperature fuel cells. In [120] mechanical activation was applied to preparation of high dispersed $LaCoO_3$ with better catalytic activity. The mixtures of anhydrous oxides ($La_2O_3$ and CoO) and hydrated compounds ($La_2(CO_3)_3{\cdot}6H_2O$ and $CoCO_3{\cdot}Co(OH)_2$) were activated for 5 min in high energetic planetary mill and then heated at 700°C for 2 h. It was shown that in nonactivated mixtures, no noticeable interaction occurs at chosen conditions, while a partial reaction is observed in the activated mixtures of anhydrous oxides. Moreover, the interaction in activated mixtures of hydrated compounds is completed even at lower temperatures. After heating at 500°C, the product is amorphous; crystallization occurs above 550°C. The specific surface area of this product goes through a maximum at 550°C (12 $m^2/g$). The increase of temperature up to 700°C brings to an increase of crystallites' size and, as a consequence, to decrease of specific surface area.

The catalytic activity of samples as prepared appeared to be higher than the activity of samples prepared by ceramic method [121]. The specific catalytic activity in CO and butane was found to be a function of the calcination temperature with its maximum at 700°C. Probably, the reason consists in the differences of chemical composition and the structure of the surface. SIMS data have demonstrated that middle- and high-temperature samples differ by the transition metal surface concentration. In some cases, high level of activity of middle-temperature defective samples can be assigned to segregation of Co ions on the surface of their particles.

## 6.8. Indium tin oxide

Tin doped $In_2O_3$ (ITO) films, widely used in photonic devices, are prepared by sputtering [122-127]. The sputtering technique requires sufficiently dense targets without any additives for highly conductive films [128-130]. Due to low sinterability of ITO [131,132], however, a costly hot-pressing procedure is often necessary. Another application of soft mechanochemistry where dissolution – reprecipitation is involved was proposed [133].

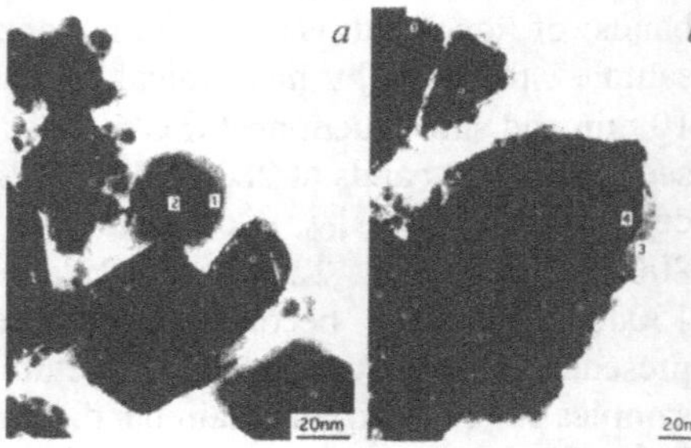


*Figure 6.39*. TEM images of the sample IS-P-Z120 (*a*) and IS-NP (*b*).

Mechanical activation is less expensive alternative to increase sinterability of ITO, if contamination is suppressed within a tolerable level. Composite powder prepared by an in situ sol method using $SiO_2$ beads as grinding media was densified up to 94% of the theoretical density after heating at temperatures as low as 1300°C [133]. However, the influence of silica introduced from grinding media as sintering aids has not sufficiently been taken into account. Silica appreciably enhances the sinterability, but it seriously degradates the film properties. Therefore, $ZrO_2$ beads were then used. Zirconia is known to suppress densification of ITO because of its preferential segregation at the grain boundary [134]. Hence, enhanced sinterability, if any, cannot be attributed to the chemical effect of $ZrO_2$.

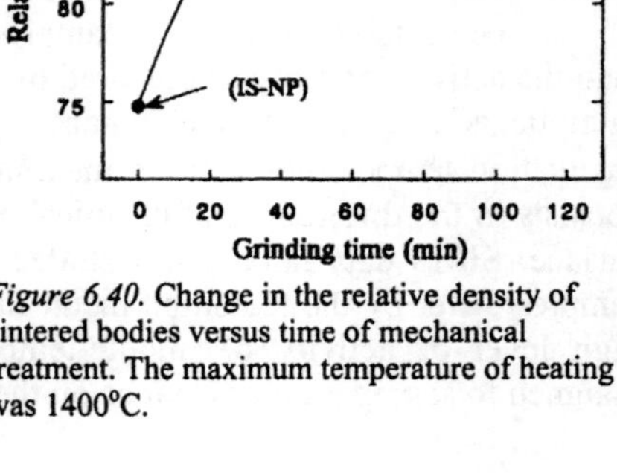


*Figure 6.40*. Change in the relative density of sintered bodies versus time of mechanical treatment. The maximum temperature of heating was 1400°C.

A typical recipe is given below: 114.8 g of $In_2O_3$ fine powders were dispersed into 480 $cm^3$ of 0.1 M aqueous solution of $SnCl_4$. The slurry was digested in an agitation mill using 0.8 mm beads of $ZrO_2$, partially stabilized with $Y_2O_3$, as grinding media. After 5 min, 18.4 $cm^3$ of 11.1 M $NH_3 \cdot aq$ was added while grinding was continued up to 120 min.

After repeated centrifugation and decantation to wash out chlorine ion, the precipitate was dried in air. Poly (vinyl alcohol) was added to the precursor mixture and dried in air. The dried mass was crushed to pass a 150 mesh sieve and compressed uniaxially at 180 MPa. After the binder was burnt out, the specimen was fired in dry air at a constant heating rate, 10K·min$^{-1}$, up to 800°C. Heating was resumed at 1.7 K·min$^{-1}$ up to the maximum temperature, 1300°C or 1400°C, and held for 5 h.

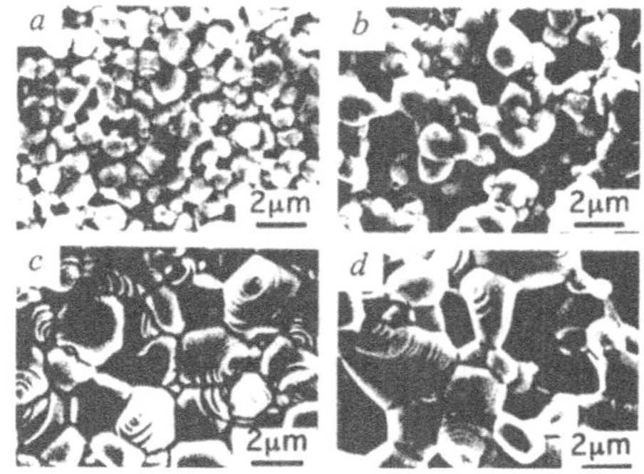


*Figure 6.41.* SEM images of the themally etched surface of the sintered body. *a*: IS-P-Z 120 and *b*: IS-NP; after heating at 1300°C; and *c*:IS-P-Z 120 and *d*: IS-NP after heating at 1400°C. Sintering time: 5h for all the samples.

Fig. 6.39 shows TEM images of the mixtures with and without grinding. While some regions on the 120 min ground sample (IS-P-Z 120) seem to be structureless without any fringe pattern (Fig. 6.39a), all the near-surface particles, including those with 1 - 2 nm in diameter, show fringe patterns on the intact mixture (Fig. 6.39b). The fringe pattern in the interior of relatively large particles corresponds to that of $In_2O_3$ (100). The surface morphology is smoother on the sample with precipitation during grinding (Fig. 6.39a). The average particle size of the near-surface small particles are correspondingly larger.

Fig. 6.40 shows the change in the relative density of the sintered body with preliminary grinding time for the samples obtained after heating up to 1400$^{O}$C and held for 5 h. The relative density increased rapidly with grinding time and reached 91% of the theoretical density after grinding for 120 min. Fine structure is significantly changed under mechanical activation. Fig. 6.41 shows that steady densification took place on the sample of in situ precipitation during grinding

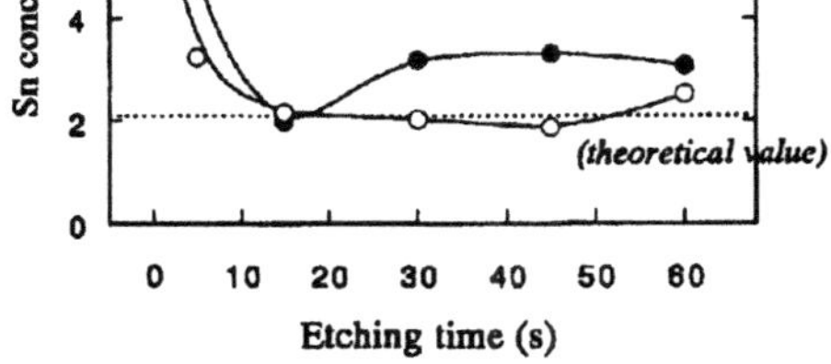


*Figure 6.42.* Depth profile of Sn for the green compacts from IS-P-Z 120 and IS-NP.

for 120 min (a and c). In contrast, large pores and chain-like particles are observed for the samples without grinding (b and d).

As shown in Fig. 6.42, Sn concentration decreased similarly after etching by $Ar^+$ ion beam, regardless of grinding during precipitation, as far as the green compacts were concerned. The Sn concentration profile of the samples after sintering of the non-ground mixture was quite different (Fig. 6.43), i.e., high Sn concentration extends even after etching for more than 240 s. The Sn concentration of the non-milled sample (IS-N) after 240 s etching was still more than twice as large as that of the milled sample (IS-P-Z 120).

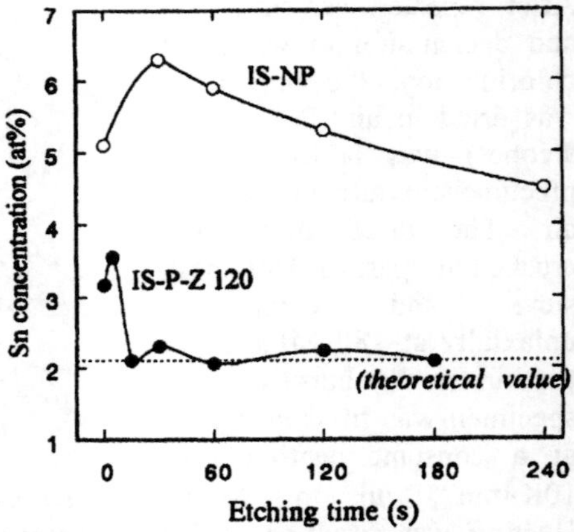


*Figure 6.43.* Depth profile of Sn for the sintered bodies from IS P-Z 120 and IS-NP.

As shown in Fig. 6.44*a*, the weight loss increased with increasing grinding time, accompanied by an exothermic DTA peak at 320°C, as shown in Fig. 6.44*b*. Due to dissolution of the outer layer of $In_2O_3$ under mechanical stress and coprecipitation with stannic ion, a hydrogel, $xIn_2O_3 \cdot ySnO_2 \cdot nH_2O$ was formed. Enhanced dissolution - precipitation results in the smoother surface and larger area of newly-born area without showing any fringe pattern even on a high-resolution TEM image.

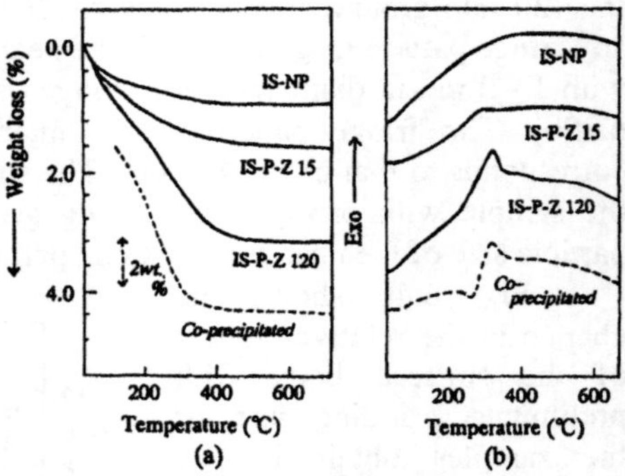


*Figure 6.44.* TG (a) and DTA (b) profiles for the samples ground for various periods and co-precipitated samples.

When the width of the channels between loosely bound grains in the sintered body decreases, the vaporization rate decreases and the rate of capture of SnO increases. This will favor the densification of ITO. The diffusion rate of SnO gas molecules is stopped when the width of the effusion channel is smaller than the mean free path of SnO gas.

Thus, densification of ITO ceramics starting from the 'in situ' precipitation of the stannic components during grinding in an aqueous milieu is based on the formation of co-precipitated complex hydrogels around the surface of the particles due to dissolution-reprecipitation. This results in the decrease in the width of the channel for SnO effusion, and hence, the vaporization of $SnO_2$ in the form of SnO (g).

## *References*


1. Kotsupalo N.P., Menzheres L.T., Mamylova E.V., Ryabtsev A.D. The methods for the preparation of the sorbent $LiCl\cdot 2Al(OH)_3\cdot mH_2O$ to recover lithium from liquids. Chem. Sustainable Development 1999; 7; 249-59.
2. Poroshina T.A., Kotsupalo N.P., Menzheres L.T., Isupov V.P. Crystal chemical features of the anion forms of double hydroxide of aluminium and lithium. Zhurn. Strukt. Khimii 1994, 35; 158-63.
3. Kotsupalo N.P., Menzheres L.T., Mamylova E.V. Sorption properties of defect forms of $LiCl\cdot 2Al(OH)_3\cdot mH_2O$. Zhurn. Prikl. Khimii 1998; 71: 1140-45.
4. Avvakumov E.G. Soft mechanochemical synthesis – a basis for new chemical technologies. Chem. Sustainable Development 1994; 2-3: 485-98.
5. Mackenzie K.J.D., Temuujin J., Jadambaa T.S., Smith M.E., Angerer P., Mechanochemical synthesis and sintering behavior of magnesium aluminate spinel. J. Mat. Science 2000; 35: 5529-35.
6. Isupov V.P., Chupakhina L.E., Mitrofanova R.P. Mechanochemical synthesis of layered double hydroxides. In Book of Abstracts. III Int. Conf. INCOME, Prague, 3-8 September, 2000.
7. Figsch V., Burianova E. Synthesis of spinel-magnesia ceramics from finely milled mixtures of magnesite $MgCO_3$ and gibbsite $Al(OH)_3$. In Book of Abstract I Int. Conf. INCOME, Kosice, 23-26 March,1993
8. Barzakovsky V.P., Lapin V.V., Boykova L.I., Kurtseva N.N., *Diagrams of State of Silicate Systems. Issue 4. Ternary Oxide Systems.* Leningrad: Nauka, 1974.
9. Prokofyev V.Yu., Ilyin A.L., Shirokov Yu.T., et al Mechanochemical synthesis of calcium hydroaluminate. Izvestia Vuzov. Khim. I Khim. Tekhnol. 1995; 38: 24-27.
10. Filio J.M., Perioho R.V., Saito F. et al. Mechanosynthesis of tricalcium aluminium hydrate of mixed grinding. Mat. Sci. Forum 1996; 503: 225-31.
11. Kano I., Yamashita H., Saito F. Effect of heat-assisted grinding of a calcium hydroxide – gibbsite mixture on formation of hydrated calcium aluminate and its hydration behavior. Powder Technology 1998, 98; 279-80.
12. Temuujin I., Mackenzie K.I.D. et al. Effect of mechanochemical treatment on the synthesis of calcium dialuminate. J. Mat. Chem. 2000; 10: 1019-23.
13. Kvitkovsky A.K., Kosova N.V., Avvakumov E.G. et al. Mechanochemical synthesis of calcium hydroaluminates. Chem. Sustainable Development 2000; 8: 686-91.
14. Prokofyev V.Yu., Ilyin A.P., Sazanova T.V. Investigation of mechanochemical processes of the joint activation of hydrargillite and calcium compounds. Neorg. Materialy 2000; 36: 1076-81.
15. Kuznetsova T.V., Talaber I., *Alumina cement.* Moskva: Stroyizdat, 1988.
16. Yakerson V.I., Golosman E.Z., Efremov V.N. Preparation and properties of the adsorbent and support on the basis of calcium dialuminate, a component of high-alumina cement. Kinetika i Kataliz 1994; 35: 626-30.
17. Ilyin A.P., Prokofyev V.Yu., Sazanova T.V. The development of absorbers for the absorption purification of technological gases from fluorine compounds. Zhurn. Prikl. Khimii 1999; 72: 1489-92.
18. Andryushkova O.V., Kirichenko O.A., Ushakov V.A., Poluboyarov V.A. Solvent-free method of the synthesis of substituted barium hexa-aluminates. Chem. Sustainable Development 1994; 2: 451-54.
19. Krivoruchko O.P., Buyanov R.A., Paramzin S.M., Zolotovsky B.P. The interaction of mechanochemically activated hydroxides of Al(III) with crystal oxides of divalent metals. Kinetika i Kataliz 1988; 29: 252-53.
20. Zhang Q., Saito F. Mechanochemical synthesis of lanthanum aluminate by grinding lanthanum oxide with transition alumina. J. Am. Ceram. Soc. 2000; 83: 439-41.

21. Liao J., Senna M. Enhanced dehydration and amorphization of $Mg(OH)_2$ in presence of ultrafine $SiO_2$ under mechanochemical conditions. Thermochim. Acta 1992; 210: 89-102.
22. Temuujin J., Okada K., Mackenzie K.J.D. Role of water in mechanochemical reactions of MgO - $SiO_2$ systems. J. Solid State Chem. 1998; 138: 169-77.
23. Yang J.C. The system magnesia- silica-water below 300°C. I. Low temperature phases from 100 to 300°C and their properties. J. Am. Ceram. Soc. 1960; 43: 542-49.
24. Temuujin J., Okada K., Mackenzie K.J.D. Formation of layered magnesium silicate during the ageing of magnesium hydroxide - silica mixtures. J. Am. Ceram. Soc. 1998; 81: 574-76.
25. Kosova N.V., Devyatkina E.T., Avvakumov E.G. Mechanochemical synthesis of calcium silicates on the basis of hydrated oxides. Sib. Khim. Zhurn. 1992; 2: 135-43.
26. *The chemistry of cements.* Ed. Taylor H.F.W., London, New-York, 1964.
27. Mi G., Saito F., Hanada M. Mechanochemical synthesis of tobermorite by wet grinding in a planetary ball mill. Powder Technology 1997; 93: 77-81.
28. Saito F., Mi G., Hanada M. Mechanochemical synthesis of hydrated calcium silicates by room temperature grinding. Book of Abstracts. XIII Int. Symp. Reactivity of Solids. Hamburq, 8-12 Sept. 1996.
29. Sasaki K., Masuda T., Ishida H., Mitsuda T. Synthesis of calcium silicate hydrate with Ca/ Si = 2 by mechanochemical treatment. J. Am. Ceram. Soc. 1996; 80: 472-76.
30. Schneider H., Okada K.,Pask J. Mullite and mullite ceramics. J. Am. Ceram. Soc. Bull. 1977; 56: 1008-11.
31. Berry Leonard, Mason Brian, Dietrich Richard, *Mineralogy*. Second Edition, San Francisco: W.H. Freeman and Company, 1983.
32. Klevtsov D.P., Krivoruchko O.P., Zolotovsky B.P., Buyanov R.A. Study of solid-phase transformation during mechanochemical and thermal treatment of aluminosilicate systems, using thermal analysis. Thermochim. Acta 1985; 93: 513-15.
33. Temuujin J., Okada K., Mackenzie K.J.D. Formation of mullite from mechanochemically activated oxides and hydroxides. J. European Ceram. Soc. 1998; 18: 831-35.
34. Temuujin J., Jadambaa T.S., Okada K., Mackenzie K.J.D. Mechanochemical preparation of aluminosilicate precursors from gibbcite - silica acid mixtures. Mat. Letters 1998; 36: 48-51.
35. Temuujin J., Okada K., Mackenzie K.J.D. Characterization of aluminosilicate (mullite) precursors prepared by mechanochemical process. J. Mat. Res. 1998; 13: 2184-88.
36. Temuujin J., Jadambaa T.S., Okada K., Mackenzie K.J.D. Preparation of aluminosilicate precursor by silica mixtures. Bull. Mat. Sci. 1998; 21: 185-87.
37. Karagedov G.R.,Liubuschko G.I. Mechanochemically stimulated synthesis of monophase mullite. Chem. Sustainable Development 1998; 6: 161-63.
38. Kawai S., Yoshida M., Hashizume G. Preparation of mullite from kaolin by dry-grinding. J. Ceram. Soc. Japan 1990; 98: 669-74.
39. Klevtsov D.P., Mastikhin V.M., Krivoruchko O.P., et al. Study of solid-state transformation mechanism on thermal treatmemt of alumosilicate systems. II. Influence of mechanical treatment on kaolinite state and following transformations under calcination. Izvestia SO AN SSSR, ser. khim.nauk 1988; 3 : 62-69.
40. Temuujin J., Mackenzie K.J.D., Schmucker M., et al. Phase evolution in mechanically treated mixtures of kaolinite and alumina hydrates (gibbsite and boehmite). J. Europe Ceram. Soc. 2000; 20: 413-21.
41. Matveeva F.A., Plekhanova E.A., *Physiko-Chemical Investigations of Alumosilicates and Zirconiumcontaining Systems and Materials.* Novosibirsk: Nauka, 1972.
42. Avvakumov E.G., Chizhevskaya S.V., Stoyanov E.S., et al. Influence of components nature of mechanically activated mixture of zirconium and silicium oxides on solid-state synthesis of zircon. Zhurn. Prikl. Khimii 1999; 72: 1420-24.

43. Cao J., Li Z. Mechanically activated synthesis and application of ultrafine $ZrSiO_4$ powder. Trans. Nonferrous Metal. Soc. China 1998; 8: 259-62.
44. Shin Y.-C., Shin D.-Y., Yans S.-M. , et al. Synthesis $ZrSiO_4$ powders by the sol-gel process. Effect of milling. Yoop Hakhoechi 1995; 32: 853-57.
45. Baek J. G., Isobe T., Senna M. Mechanochemical effects on the precursor formation and microwave dielectric characteristics of $MgTiO_3$. Solid State Jonics 1996; 90: 269-79.
46. Kosova N.V., Avvakumov E.G. Mechanochemical synthesis of calcium titanate on the basis of hydrated oxides. Sib. Khim. Zhurn. 1992; 3: 121-27.
47. Pletnev R.I., Ivashin A.A., Kleshchev D.G., *Hydrated Oxides of the Elements of IV and V Groups*. Moskva: Nauka, 1986.
48. Biryuk L.I., Goroshenko L.G., Khondros E.L., Kalinichenko A.M. Change of composition and structure of titanium hydroxides depending on heating. Ukrainskii Khim. Zhurn. 1971; 37: 1221-24.
49. Kosova N.V., Devyatkina E.T., Avvakumov E.G., et al. Mechanochemical reactions of hydrated oxides: some aspects of mechanism. Chem. Sustainable Development 1998; 6: 125-30.
50. Kosova N.V., Devyatkina E.T., Denisova T.A. et al. Proton transfer in mechanochemical reactions of hydrated oxides. Zhurn. Neorg. Khimii 1999; 44: 912-16.
51. Kosova N.V., Asanov I.P., Devyatkina E.T. et al. Investigation of atom state change in surface layer at the initial stages of mechanochemical reactions by means of X-ray electron microscopy. Zhurn. Neorg. Khimii 1999; 44: 100-2.
52. Kosova N.V., Avvakumov E.G., Malakhov V.V. et al. About the nature of phases formed in soft mechanochemical synthesis of calcium titanate. Dokl. Akad. Nauk 1997; 356: 350-53.
53. Avvakumov E.G., Sysoev B.F., Kosova N.V. The role of hydrolytic interactions in solid-phase mechanical synthesis of barium titanate. Proc. of All-Union Conf. "Mechanochemical Synthesis". Vladivostok, August 20 – 23, 1990.
54. Sysoev V.F., Mali V.I. Chemical processes in the systems $BaO_2 - TiO_2$ and $PbO - TiO_2$ under shock compression. Neorg. Materialy 1993; 29: 83-87.
55. Thomas H.A.J., Stevens R. Aluminium titanate. A literature review. Br. Ceram. Trans. J. 1989; 88: 4-229.
56. Freudenberg B., Mocellin A. Aluminium titanate formation by solid state reaction of fine $Al_2O_3$ and $TiO_2$ powders. J. Am. Ceram Soc. 1987; 70: 33-38.
57. Kojima Y., Isobe T., Senna M. Mechanisms of Al and titania hydrogel complex formation via a mechanical route. J. Mater. Res. 1996; 11: 1305 –9.
58. Kojima Y., Isobe T., Senna M., et al. An XAFS study on reconstruction of short-range order in mechanically alloyed Al-Ti and Al-Ti-O complexes. J. Alloys and Compounds 1997; 248: 42-58.
59. Kupin A.V., Prokofyev V.Yu., Ilyin A.P. Synthesis of aluminium titanate using stabilizing additives. Steklo i Keramika 1999; 4: 20-23.
60. Karakchiev L.G., Avvakumov E.G., Vinokurova O.B., Gusev A.A., Zima T.M., Lyakhov N.Z. Comparison of sol-gel and mechanochemical methods of preparation of $Al_2TiO_5$. Chem. Sustainable Development 2001; 9: 27-34.
61. Kojima Y., Jsobe T., Senna M., et al. Mechanochemical synthesis of $Al_5TiO_5$ J. Mechanochemistry and Mechanical Alloying 1994; 1: 34-39.
62. Kojima Y., Senna M., Shinohara T.,et. al. Mechanochemical synthesis $Al_5TiO_5$. J. Alloys and Compounds 1995; 227: 97-104.
63. Shlyahtina A.V., Shcherbakova L. G., Kolbanev J. V., Minaeva N. A. Syntesis of rate-earth dititanates using fine-grained reagents. Book of Abstracts. III Int. Conf. INCOME, Prague, September 4-8, 2000.
64. Christoffersen R., Danes P.K. Structure of commensurate and incommensurate ordered phases in system $ZrTiO_4 - Zr_5Ti_7O_{24}$ . J. Amer. Ceram. Soc. 1992; 75: 563-69.

65. Bordet P., McHall A., Santoro A., Roth R. S. Powder neutron diffraction study $ZrTiO_4$, $Zr_5Ti_7O_{24}$ and $FeNb_2O_6$. J. Solid State Chem. 1986; 64: 30-46.
66. Azough F., Wright A., Freer R. The microstructure and dielectric properties of $Zr_5Ti_7O_{24}$ ceramics. J. Solid State Chem. 1994; 108: 284 – 290.
67. Fotiev A.A., Volkov V.L., Kapustkin V.K., *Oxide vanadium bronzes.* Moskva: Nauka; 1978.
68. Molchanov V.V., Maksimov G.M., Plyasova L.M., Goidin V.V., Kozhevnikov I.V. Mechanochemical synthesis of vanadates of alkali metals. Neorg. Materialy 1993; 29: 555-58.
69. Wadsley A.D. Crystal chemistry of non-stoichiometric pentavalent vanadium oxides: crystal structure of $Li_{1+x}V_3O_8$. Acta Cryst. 1957; 10: 261-67.
70. Wickham D.G. A study of the solid solutions $Li_{1+x}V_3O_{8\pm\lambda}$ and the preparation of $LiVO_3$ and $Li_3VO_4$. Inorg. Nucl. Chem. 1965; 27: 1939-46.
71. Panero S., Pasquali M., Pistoia G. J. Rechargeable Li/Li $_{1+x}V_3O_8$ cells. Electrochem. Soc. 1983; 130: 1225-26.
72. Pistoia G., Pasquali M., Wang G., Li L. Li/$Li_{1+x}V_3O_8$ secondary batteries. Synthesis and characterization of an amophous form of the cathode. J. Electrochem. Soc. 1990; 137: 2365-70.
73. Dai J., Li S.F.Y., Gao Z., Siow K.S. Low-temperature synthesized $LiV_3O_8$ as a cathode material for rechrgeable lithium batteries. J. Electrochem. Soc. 1998; 145: 3057-62.
74. Kosova N.V., Devyatkina E.T., Kozlova S.G., Anufrienko V.F., Vosel S.V., Vasenin N.T. Low temperature synthesis of $LiV_3O_8$ by mechanical activation. 5 Journees du Reseau Francais de Mechanosynthese, Bordeaux, May 23-24, 2000.
75. Kosova N.V., Anufrienko V.F., Vasenin N.T., Vosel S.V., Devyatkina E.T. Peculiarities of $Li_{1+x}V_3O_8$ structure according to EPR spectrocsopy data. J. Solid State Chem. 2001 (to be published).
76. Kosova N.V., Vosel S.V., Anufrienko V.F., Vasenin N.T., Devyatkina E.T. Reduction processes in the course of mechanochemical synthesis of $Li_{1+x}V_3O_8$. J. Solid State Chem. 2001 (to be published).
77. Kera Y. Infrared study of alkali tri- and hexavanadates as formed from their melts. J. Solid State Chem. 1984; 51: 205-11.
78. Avvakumov E.G., Anufrienko V.F., Vosel S.V., Gadzhieva F.S., Kalinina N.G., Polyuboyarov V.A. Investigations on structural changes in mechanically activated titanium and vanadium oxides by EPR. Izvestiya SO AN SSSR, seriya khim. nauk 1987; 1: 41-8.
79. Ioffe V.A. and Patrina I.B. On the state of carriers in $V_2O_5$ according EPR. Solid State Phisics. 1968; 10: 815-21.
80. Hasegawa H. Dynamic properties of s-d intercalation. Progress of Theoretical Physics. 1959; 21: 483-500.
81. Hirst L.L., Schafer W., Seipler F., Elschner B. Doubly bottlenecked EPR in chromium, manganese-doped copper oxide. Phys. Rev. B. 1973; 8: 64-69.
82. Slobodin B.V., Zhiljaev V.A., Fotiev A.A. Arapova I.A., Tugova N.P. A thermoanalytical study of the intercalation of vanadium-oxide (V) with calcium oxide and calcium carbonate. J. Therm. Anal. 1979; 15: 197-206.
83. Goni S., Sobrados L., Hernandez M.S. Increase of acid-surface reactivity through water molecules absorption process: $V_2O_5$-$CaCO_3$ behaviour. Solid State Ionics. 1993; 63-65: 786-90.
84. Kosova N.V. and Avvakumov E.G. Mechanochemical synthesis of calcium vanadates. Sib. Khim. Zhurn. 1991; 6: 128-35.
85. Lee W.-S., Isobe T., Senna M. Formation of $SrVO_3$ precursors from $Sr(OH)_2$ and $V_mO_n$ ($2/5 \leq m/n \leq 2/3$) by mechanical stressing under ambient or reducing atmosphere. Solid State Ionics. 1999; 123: 293-300.

86. Sobalik Z., Lapina O.B., Novgorodova O.N., Mastikhin V.M. Interaction of vanadia with alumina and titania during ultra-high intensity grinding at room temperature as evidenced from $^{51}V$ NMR spectra. Applied Catalysis. 1990; 63:191-95.
87. Vejux A., Courtine P. Interfacial reactions betwee $V_2O_5$ and $TiO_2$ (anatase): role of the structural properties. J. Solid State Chem. 1978; 23: 93-103.
88. Molchanov V.V., Plyasova L.M., Goidin V.V., Lapina O.B., Zaikovskii V.I. New compounds in $MoO_3$-$V_2O_5$ system. Neorg. Materialy 1995; 31: 1225-29.
89. Shanha K., Subbanna G.N., Varma K.B.R. Mechanically activated synthesis of nanocrystalline powders of ferroelectric vismuth vanadate. J. Solid State Chem. 1999; 142: 41-7.
90. Nivoix V., Bernard F., Gaffet E., Perriat P., Gillot B. Mechanical activation conditions of the $Fe_2O_3$ and $V_2O_3$ mixture powders in order to obtain a nanometric vanadium spinel ferrite. Powder Technol. 1999; 105: 155-61.
91. Tarascon J.M., Wang E., Shokoohi F.K., McKinnon W.R., Colson S. The spinel phase of $LiMn_2O_4$ as a cathode in secondary lithium cells. J. Electrochem. Soc. 1991; 138: 2859-68.
92. Thackeray M.M. Spinel electrodes for lithium batteries. J. Am. Ceram. Soc. 1999; 82: 3347-54.
93. Thackeray M.M., de Kock A., David W.I.F. Synthesis and structural characterization of defect spinels in the lithium-manganese-oxide system. Mat. Res. Bull. 1993; 28: 1041-49.
94. Kosova N.V., Uvarov N.F., Devyatkina E.T., Avvakumov E.G. Mechanochemical synthesis of $LiMn_2O_4$ cahtode material for lithium batteries. Solid State Ionics. 2000; 135: 107-14.
95. Kosova N.V. Mechanochemical synthesis of cathode materials for lithium batteries. Materials for lithium-ion batteries. NATO Science Series. 3. High Technology. 2000; 85: 497-500.
96. Kosova N.V., Devyatkina E.T., Kozlova S.G. Mechanochemical way for preparation of disordered lithium-manganese spinel compounds. Proceedings of the 10$^{th}$ Int. Meet. on Lithium Batteries "Lithium 2000"; Como, May 28-June 2, 2000; J. Power Sources 2001 (to be published).
97. Kosova N.V., Kozlova S.G., Gabuda S.P., Avvakumov E.G. Spin density transfer through phase contact boundary in the products of mechanical activation. Doklady AN 1998; 362: 493-96.
98. Kosova N.V., Asanov I.P., Devyatkina E.T., Avvakumov E.G. State of manganese atoms during the mechanochemical synthesis of $LiMn_2O_4$. J. Solid State Chem. 1999; 146: 184-88.
99. Gao Y. And Dahn J.R. Thermogravimetric analysis to determine the lithium to manganese ratio in $Li_{1+x}Mn_{2-x}O_4$. Appl. Phys. Lett. 1995; 66: 2487-89.
100. Isupova L.A., Sadykov V.A., Avvakumov E.G., Kosova N.V. Mechanochemical activation in technology of high-temperature oxide catalysts. Chem. Sustainable Development 1998; 6: 199-202.
101. Fernandez-Rodriguez J.M., Morales J., Tirado J.L. Synthesis and alteration of $\alpha$-$LiFeO_2$ by mechanochemical processes. J. Mat. Science. 1988; 23:2971-74.
102. Karagedov G.R., Konovalova E.A., Startzev S.I., Lyakhov N.Z. Mechanochemical synthesis of lithium ferrites. Proc. All-Union Conf. "Mechanochemical synthesis", Vladivostok, 1990.
103. Karagedov G.R., Konovalova E.A., Boldyrev V.V., Lyakhov N.Z. Influence of reagent biography and reaction conditions on kinetics of lithium ferrite synthesis. Solid State Ionics. 1990; 42: 147-51.
104. Rabanal M.E., Varez A., Levenfeld B., Torralla J.M. Synthesis of spinel $MgFe_2O_4$ by mechanical grinding. Proc. VII European Conf. on Solid State Chem. Madrid, Sept. 15-18, 1999.

105. Savinkina M.A., Zyryanova V.N., Logvinenko A.T., Voronin A.P. Influence of mechanical activation on the synthesis anf chemical activity of dicalcium ferrite. Izv. SO AN SSSR, ser. khim. nauk 1983; 5: 85-7.
106. Kosova N.V., Devyatkina E.T., Avvakumov E.G., Gainutdinov I.I., Rogachev A.Yu., Pavlyukhin Yu.T., Isupova L.A., Sadykov V.A. Mechanochemical synthesis of dicalcium ferrite with the perovskite structure. Neorg. Materialy. 1998; 34: 385-90.
107. Schmidt M. And Kaczmarek W.A. Synthesis of $SrFeO_{2.5}$ from mechanically activated reactants. J. Alloys and Compounds. 1999; 283: 117-21.
108. Schapkin V.L., Kormiliztina Z.A., Avvakumov E.G. Mechanochemical synthesis of amorphous and crystalline barium ferrites. Proc. All-Union Conf. "Mechanochemical Synthesis", Vladivostok, 1990.
109. Mizushima K., Jones P.C., Wiseman P.J. and Goodenough J.B. $Li_xCoO_2$ ($0<x\leq1$): a new cathode material for batteries of high energy density. Mater. Res. Bull. 1980; 15: 783-89.
110. Orman H.J. and Wiseman P.J. Cobalt lithium oxide, $CoLiO_2$: structure refinement by powder neutron diffraction. Acta Crystallogr. C. 1984; 40: 12-14.
111. Gummow R.J., Liles D.C., Thackeray M.M. and David W.I.F. A reinvestigation of the structures of lithium-cobalt-oxides with neutron-diffraction data. Mat. Res. Bull. 1993; 28: 1177-84.
112. Garcia B., Farcy J., Pereira-Ramos J.P. and Baffier N. J. Electrochemical properties of low-temperature crystallized $LiCoO_2$. Electrochem. Soc. 1997; 144: 1179-84.
113. Tarascon J.M., Wang E., Shokoohi F.K. et al. The spinel phase of $LiMn_2O_4$ as a cathode in secondary lithium cells. J. Electrochem. Soc. 1991; 138: 2859-64.
114. Chiang Y.-M., Jang Y.-I., Wang H., Huang B., Sadoway D.R. and Ye P. Synthesis of $LiCoO_2$ by decomposition and intercalation of hydroxides. J. Electrochem. Soc. 1998; 145: 887-91.
115. Amatucci G.G., Tarascon J.M., Larcher D. and Klein L.C. Synthesis of electrochemically active $LiCoO_2$ and $LiNiO_2$ at 100°C. Solid State Ionics. 1996; 84: 169-80.
116. Fernandez-Rodriguez J.M., Morales J. and Tirado J.L. Mechanochemical preparation and degradation of $LiCoO_2$. Reactivity of Solids. 1987; 4: 163-71.
117. Obrovac M.N., Mao O. and Dahn J.R. Structure and electrochemistry of $LiMO_2$ (M=Ti, Mn, Fe, Co, Ni) prepared by mechanochemical synthesis. Solid State Ionics. 1998; 112: 9-19.
118. You H.W., Lee H.-Y., Jang S.-W., Shin K.-C., Lee J.-K., Lee S.-J., Balk H.-K. and Rhee D.-S. Synthesis of $LiCoO_2$ by mechanical alloying. J. Mater. Sc. Lett. 1998; 17: 931-34.
119. Kosova N.V., Anufrienko V.F., Larina T.V., Devyatkina E.T. The synthesis of $LiCoO_2$ – cathode material for lthium-ion batteries by mechanical activation. Chem. Sustainable Development. 2001; 2: 200-206.
120. Pauli I.A., Avvakumov E.G., Isupova L.A., Poluboyarov V.A. and Sadykov V.A. the influence of mechanical activation on the synthesis and catalytic activity of lanthanium cobaltite. Sib. Chim. Zhurn. 1992; 3: 133-7.
121. Isupova L.A., Sadykov V.A., Avvakumov E.G. and Kosova N.V. Mechanochemical activation in technology of high-temperature oxide catalysts. Chem. Sustainable Development. 1998; 6:199-202.
122. Chiou B.S., Hsieh S.-T., Wu W.-F. Deposition of indium tin oxide films on acrylic substrates by radio-frequency magnetron sputtering: J. Am. Ceram. Soc. 1994; 77: 1740-44.
123. Lee S. B., Pincenti J. C., Cocco A. and Naylor D.L. Electric and optical properties of room temperature sputter deposited indium oxide: J. Vac. Soc. Technol. 1993; A11: 5-10.

124. Ishibashi S., Higuchi Y., Ota Y. and Nakamura K. Low resistivity indium tin oxide transparent conductive films. I. Effect of introducing $H_2O$ gas of $H_2$ gas during direct current magnetron sputtering. J. Vac. Soc. Technol. 1989; A8: 3-6.
125. Yasui H, Tsuda Y. ITO thin films preparated by magnetron sputtering method using ITO target (effects of plasma conditions and substrate temperature on ITO film properties). Nihon-Kikaigakkai-Ronbunsyu 1993; 59: 593-98.
126. Yao L., Hao S. and Wilkinson J.S. Indium tin oxide films by sequential evaporation. Thin Solid Films 1990; 189: 227-33.
127. Pommier R., Gril C., Maruchi J. Sprayed films of indium tin oxide and fluorine-doped tin oxide of large surface area. Thin Solid Films 1981; 97: 91-7.
128. Furusaki T., Takahashi J., Kodaira K. Preparation of ITO films by Sol-Gel method. J. Ceram. Soc. Japan 1994; 102: 200-5.
129. Vasant Kumar C.V.R., Mansingh A. Effect of target-substrate distance on the growth and properties of Rf-sputtered indium tin oxide. J. Appl. Phys. 1988; 65: 1270-80.
130. Lee C.S.B., Pincenti J.C., Cocco A., Naylor D.L. Electonic and optical properties of room temperature sputter deposited indium tin oxide. J. Vac. Sci. Technol. 1993; A11: 2742-46.
131. Nade T., Kishi H. Fabrication and characteristic of the ITO sputtering targets. Hyoumen-gizyutsu 1992; 43: 20-5.
132. Gehman B.L. Influence of Manufacturing process of indium tin oxide sputtering targets on sputtering behavior. Thin Solid Films 1992; 220: 330-36.
133. Yoshimura R., Ogawa N., Mouri T. Studies on sputtering characteristics of ITO target(3). J. TOSOH Res. 1992; 36(2): 153-58.
134. Nadaud N., Nanot M., Boch P. Sintering and electrical properties of titania- and zirconia- containing $In_2O_3$-$SnO_2$ (ITO) ceramics. J. Am. Ceram. Soc. 1994; 77: 843-46.
135. Osipyan V.G., Savchenko L.M., Avvakumov E.G., Uvarov N.F. Crystallochemistry, synthesis and properties of ordinary and mixed-layered bismuth compounds. In *Mechanochemical Synthesis in Inorganic Chemistry,* ed. Avvakumov E.G. Novosibirsk: Nauka, 1991.

# Chapter 7

# MECHANOCHEMICAL SYNTHESIS OF MULTICOMPONENT OXIDE COMPOUNDS

## 7.1. Aluminosilicates of alkaline earths

Aluminosilicates of alkaline earths are widely used as initial raw material in ceramic materials science. They are the components of feldspars. Their structure consists of chains composed of four-membered rings of $SiO_4$ and $AlO_4$ tetrahedrons elongated along C axis. These chains are united to form a frame with the sites incorporating alkaline earth atoms with coordination numbers 8 (Ca) or 9 (Ba). Anorthite, $CaAl_2Si_2O_8$, is one of well known examples of such alumosilicates [1]. The structure of magnesium aluminosilicate (cordierite), $Mg_2Al_4Si_5O_{18}$, is close to mentioned aluminosilicates. It is similar to that of beryl. The main motif of the frame structure of beryl is hexamerous ring of $SiO_4$ tetrahedrons bound to each other to form hexagonal columns. In cordierite, hexamerous rings are composed of $SiO_4$ tetrahedrons and two $AlO_4$ tetrahedrons ($Al_2Si_4O_{18}$); they are bound to $AlO_4$, $SiO_4$ tetrahedrons and $MgO_6$ octahedrons. Large cations ($Ca^{2+}$, $Na^+$, $K^+$) and molecules ($CO_2$ and $H_2O$) can occupy the centres of hexagonal columns [1].

From the economical viewpoint, the known methods of the preparation of these compounds are insufficiently effective. As a rule, they are synthesized by solid-phase methods from ordinary or complex oxides by heating the mixtures at high temperatures (1100 - 1400°C) for a long time (100–185 h).

**Cordierite.** This material possesses a rare combination of properties, i.e., relatively high strength and hardness, low thermal expansion coefficient, high heat conductivity and plasticity; it conserves mechanical strength under thermal shocks. Therefore, its application areas are varied.

Cordierite synthesis method based on mechanical activation of mixtures of hydrated oxides of calcium, aluminium and silicon, as well as natural hydrated compounds (talc, kaolinite and gibbsite), has been developed in [2, 3]. Mechanical activation of these mixtures does not lead to the formation of new phases but provides good mixing at the cluster level giving aggregates that form cordierite during the subsequent thermal treatment.

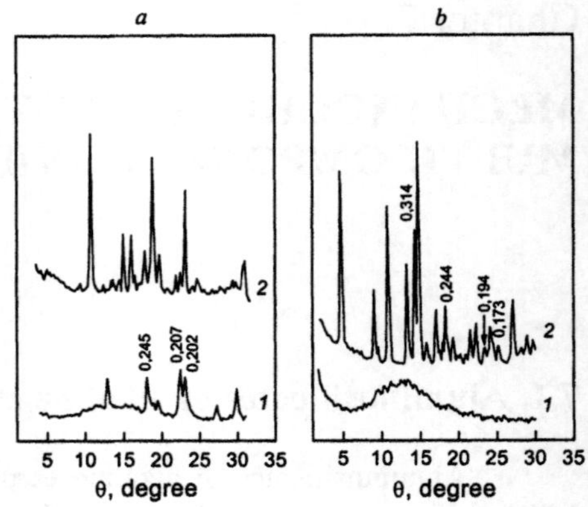


*Figure 7.1.* XRD profiles of the mixtures after mechanical activation (1) and thermal treatment at 1260°C for 2 h (2):
*a* – $MgO+Al_2O_3+SiO_2$,
*b* – $Mg(OH)_2+Al(OH)_3+SiO_2aq$.

Fig. 7.1 shows X-ray diffraction patterns of a mixture of anhydrous magnesium, aluminium and silicon oxides and a mixture of correspondent hydroxides after mechanical activation in a planetary mill followed by annealing at 1260°C for 2 h. In the former case, the reflections of spinel $MgAl_2O_4$ (0.245, 0.207, 0.202 nm) are observed in the patterns. When the mixture of hydroxides is used, the diffraction patterns contain the reflections of magnesium silicates: enstatite $MgSiO_3$ (0.314, 0.194 nm) and forsterite $Mg_2SiO_4$ (0.173, 0.244 nm), besides the reflections of the final product (cordierite). This means that the reaction in the mixtures of hydrated oxides follows the route different from that in the mixture of anhydrous oxides. At first, magnesium and aluminium hydroxides interact with each other to give double oxides, the latter reacting with alumosilicate to form cordierite:

$$2\,Mg(OH)_2 + 4Al(OH)_3 + 5SiO_2{\cdot}nH_2O \rightarrow$$
$$MgSiO_3 + Mg_2SiO_4 + 2(mAl_2O_3{\cdot}nSiO_2{\cdot}H_2O) + x\,H_2O \rightarrow$$
$$Mg_2Al_4Si_5O_{18} + y\,H_2O,$$

In this case the stage of spinel formation is absent.

In industry, cordierite is usually obtained by calcination of the mixtures containing talc, kaolinite and silica at 1300-1450°C for 20-60 h. The product contains the impurity phases: spinel, mullite, clinoenstatite, etc., that worsen the exploitation characteristics of cordierite. Since the mentioned minerals contain structural water, chemical interaction between them during mechanical activation can be considered from the viewpoint of soft mechanochemical synthesis. Mechanical activation of this mixture does simplifies the interaction between its components. It is sufficient to heat this mixture for 2 h at a temperature of 1260°C to obtain practically homogeneous cordierite without impurity phases (Fig. 7.2) [2-9].

The synthesized cordierite as prepared was then ground in a planetary mill for 1 min. The ceramics made of this cordierite sintered at

1340°C is characterized by a density of 2.3 g/cm$^3$, a porosity of 5-8 %, a compression strength of 230-240 MPa [8,9].

**Anorthite.** Anorthite is one of the components used for the preparation of porcelain and faience.

According to data [10,11], practically monophase anorthite can be synthesized after mechanical treatment of a mixture of calcium oxide with aluminium hydroxide and hydrated silica followed by thermal treatment at 1000°C for 4 h. Fig. 7.3 shows X-ray pattern of this mixture. After activation for 5 min (Fig. 7.3, curve 1), the reaction mixture exhibits a small amount of individual crystal calcium oxide (0.240 nm) and aluminium hydroxide (0.482, 0.436, 0.245 nm), while the major part of CaO is bound into an X-ray amorphous product. The annealing of the activated mixture at 1000°C gives anorthite (Fig. 7.3, curve 3) while the thermal treatment of non-activated mixture brings to the formation of intermediate product, wollastonite (Fig. 7.3, curve 2). The lines of this product are absent in X-ray patterns of the activated mixture.

**Celsian.** One of the compounds important for ceramic materials science is monoclinic celsian, $BaAl_2Si_2O_8$, a compound with a feldspar structure. However, its formation during solid phase synthesis occurs in several stages with the formation of a very stable hexagonal modification of $BaAl_2Si_2O_8$ as an intermediate phase. Its subsequent transition into thermodynamically stable monoclinic celsian is hindered and requires calcination at elevated temperatures (~1600°C) either the hydrothermal treatment, or the use of mineralizers [12-14].

Mechanical activation of the mixture of initial anhydrous oxides BaO,

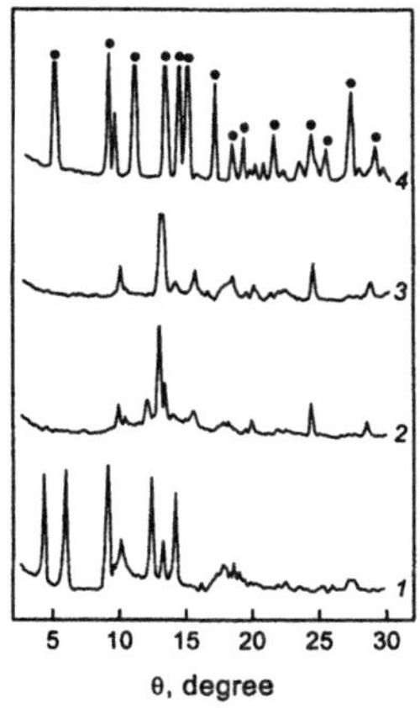


*Figure 7.2.* X-ray patterns of the mixture of talk, kaolinite and gibbsite:
1 – after mechanical activation, after the following thermal treatment for 2 h: 900 (2), 1100 (3) and 1260°C (4).
● - cordierite.

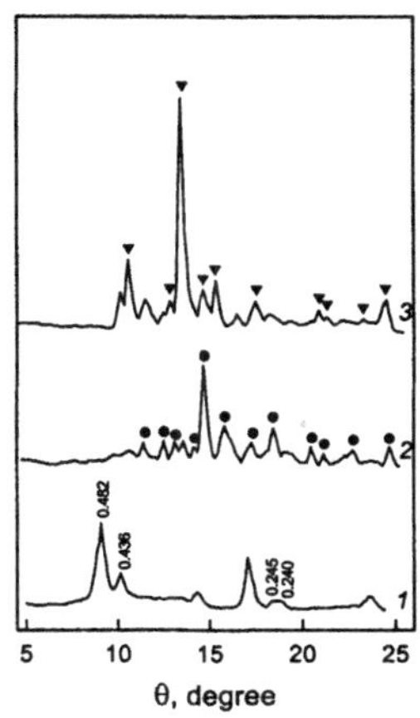


*Figure 7.3.* X-ray patterns of $Ca(OH)_2$+$Al(OH)_3$+$SiO_2$aq mixture:
1 – activated for 5 min;
2 – non-activated and annealed at 1000°C for 4 h.;
3 – activated and annealed at 1000°C.
● - wollastonite, ▼ - anorthite.

$Al_2O_3$ and $SiO_2$ with added mineralizers (KF, LiCl) provides substantial intensification of the synthesis of barium aluminosilicate. The formation of the monoclinic modification of $BaAl_2Si_2O_8$ in reaction mixtures activated in the presence of modifying agent (5 % KF) is practically completed at 1100°C within 2 h, while substantially higher temperatures and treatment duration are necessary for the formation of monoclinic celsian in the case when a usual synthesis procedure is used.

Mechanical activation of the mixture containing BaO, $Al(OH)_3$, $SiO_2 \cdot 0.69H_2O$ also intensifies the synthesis of celsian [11]. Practically monophase hexagonal celsian was obtained by mechanical activation of the mixture for 10 min and thermal treatment at 1200°C for 3 h (Fig. 7.4). Hexagonal celsian can be transformed into monoclinic modification by mechanical activation of the mixture with modifiers (KF, LiCl at a level of 1-5 mol. %) followed by annealing at 1100°C.

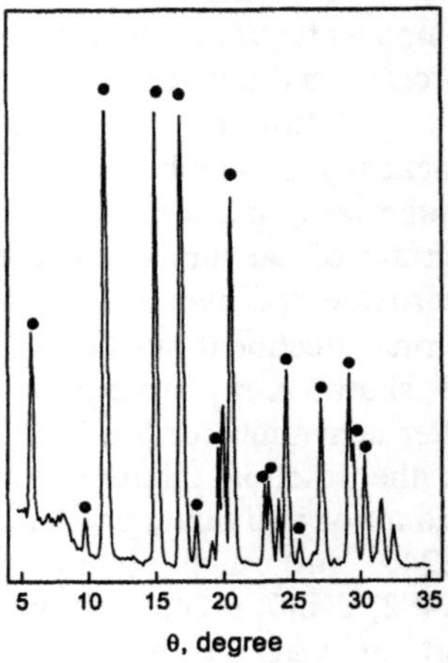


*Figure 7.4.* XRD profiles of $BaO+Al(OH)_3+2SiO_2aq$ mixture after mechanical activation and annealing at 1200°C for 2 h.
● - celsian.

## 7.2. Complex compounds with superconductive properties

Physicochemical properties of high-temperature superconductive complex oxides are substantially dependent on synthesis conditions, in particular, on homogeneity of the mixture of finely dispersed components. Mechanical activation allows to improve the ceramic technology of their preparation, in particular, $YBa_2Cu_3O_{7-x}$ and $YBa_2Cu_4O_{8-x}$, where x and y are parameters characterizing the lack of oxygen.

The application of mechanical activation for the synthesis of the mentioned above compounds started with the activation of the mixtures of anhydrous oxides [15]. Then, barium oxide (or carbonate) in the mixture was replaced with barium nitrate [16], while copper oxide was replaced with copper hydroxide or basic carbonate $Cu_2CO_3(OH)_2$ [17]. Later, yttrium and copper nitrates were used instead of oxides, while barium hydroxide $Ba(OH)_2$ or $BaO_2$ were used as barium-containing compound [18]; finally, the mixture of oxides was activated in the presence of water added for the purpose of obtaining well-molding pastes [19]. After activation, the mixtures were annealed at increased temperature, but the synthesis temperature was substantially lower than the usual synthesis temperature (by 100-150°C).

The activation of the stoichiometric mixture of the components $Y_2O_3$, $Ba(NO_3)_2$, $Cu_2CO_3(OH)_2$ with atomic ratio Y:Ba:Cu = 1:2:3 in a planetary mill for 10 min and heating within the range 675-900°C in 25° steps for 1 h resulted, according to X-ray phase analysis, in the formation of $YBa_2Cu_3O_{7-x}$ (Fig. 7.5) [17]. One can see that in the activated samples the synthesis starts within the temperature range 675-700° C. In non-activated mixtures, the formation of the product is noticeably detected only at 850°C. The increase of the amount of phase $YBa_2Cu_3O_{7-x}$ for activated samples is linearly dependent on temperature till 800°C, when the formation of the phase is practically completed. In non-activated samples, phase formation obeys the exponential law starting from the temperature 750°C, and is completed at 930°C.

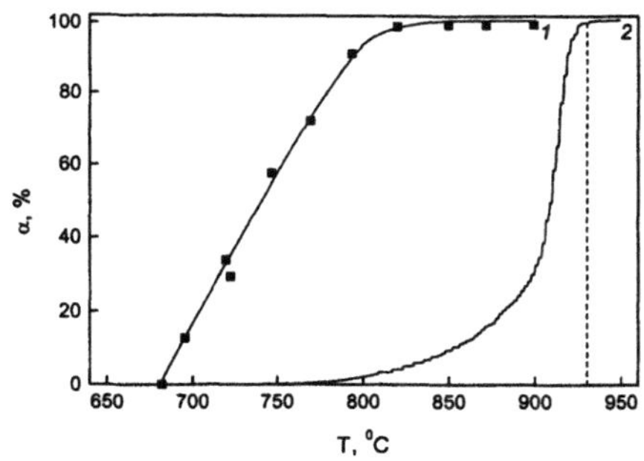


*Figure 7.5.* Formation of $YBa_2Cu_3O_{7-x}$ in $Y_2O_3$+$Ba(NO_3)_2$+$Cu_2CO_3(OH)_2$ mixture: 1 – activated, 2 – non-activated.

The synthesis of the compound $YBa_2Cu_3O_{7-x}$ and $YBa_2Cu_4O_{8-x}$ from the mechanically activated mixture of $Y(NO_3)_3$, $Cu(NO_3)_2$ and $Ba(OH)_2$ (or $BaO_2$) in the presence of small amounts of water proceeds in a similar way [18]. According X-ray phase analysis, monophase orthorhombic $YBa_2Cu_3O_{7-x}$ is already detected after annealing for 10 h at 730°C. However, its structure is not perfect yet: broad reflections in X-ray diffraction patterns are observed. Perfect orthorhombic structure is observed in materials which were thermally treated at 750°C for 5 h or at 780°C for 2 h. The studies on the formation mechanism of this compound revealed that the synthesis involves some intermediate stages, the first one being the formation of double cuprates of barium and yttrium, and the second involving the interaction between them:

$$2BaCuO_2 + 1/2Y_2Cu_2O_5 = YBa_2Cu_3O_{7-x},$$

This means that the interaction occurs via the mechanism completely excluding the formation of non-superconductive "green" phase $Y_2BaCuO_5$. Using the powder thus synthesized, one succeeded in preparing the ceramics with the density approaching the theoretical one (6.15 g/cm$^3$).

The synthesis of $YBa_2Cu_4O_{8-y}$ with lattice parameters a=0.383, b=0.387, c=2.725 nm was performed according to the same procedure involving mechanical activation of the mixture of yttrium and copper nitrates, and barium hydroxide in the presence of small amounts of water, followed by thermal treatment [18]. The synthesis is completed after heating at 720°C within 20 h. The compound $YBa_2Cu_4O_{8-y}$ is stable till 820°C, y being within the range 0.07–0.2.

In order to prepare the hardware of $YBa_2Cu_3O_{8-x}$ with complicated shapes, the mixture of $Y_2O_3$, CuO, $Ba(OH)_2$ (or $BaO_2$) is mechanically activated in the excess of water-alcohol mixture, or in alcohol. Activation results in the formation of highly concentrated suspension (paste). Along with the particles of the disperse phase, it contains a colloid component as a film gel. The paste is dried to plastic molding state and necessary humidity, then it is pressed to give a required shape and dried at 300°C for 1 h. After annealing at 900°C for 1 h, the amount of the main phase, $YBa_2Cu_3O_{8-x}$, is about 75 % when pure water is used, and ≥ 96 % when alcohol is used [19].

## 7.3. Synthesis of phase pure perovskite compounds

Metal (I) – oxygen – metal (II) bridging bond formation occurs as a consequence of a redox reaction between two metal oxides or hydroxides as mentioned above. It is particularly interesting to apply these principles to the synthesis of complex oxides such as perovskites. Phase pure perovskites, e.g. $Pb(Mg_{1/3}Nb_{2/3})O_3$(PMN), or $(1-x)Pb(Mg_{1/3}Nb_{2/3})O_3 \cdot xTiO_2$ (PMN–PT) are notorious for the difficulty to obtain as phase pure perovskite. To fabricate phase pure PMN it is necessary to prepare magnesium niobate (columbite) first, and then to heat a mixture of columbite with other oxide species. PbO is usually added in a small excess over exact stoichiometry that is a complicated and tedious process.

Since $Nb_2O_5$ is a highly acidic oxide while $Mg(OH)_2$ is a highly basic one, a milling of a stoichiometric mixture of these compounds pushes forward the formation

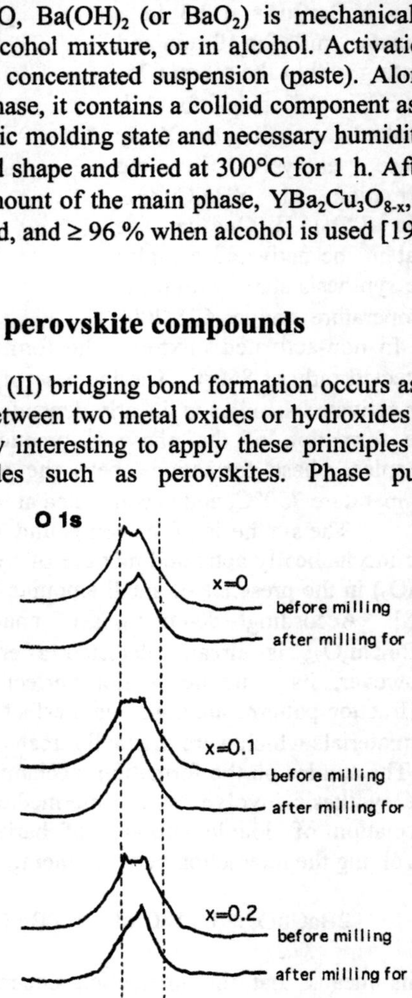


*Figure 7.6.* XPS profiles of the precursor of PMN-PT.

of hetero bridging bonds, Nb-O-Mg. Phase pure perovskites, PMN–PT (x=0~0.2), are actually obtained by heating of milled mixture as a precursor at moderate temperatures as low as 850°C [20,21]. Products are finely and homogeneously grained just because of firing at low temperature. $Mg(OH)_2$ (brucite), PbO (massicot), $TiO_2$ (anatase and rutile) and $Nb_2O_5$ (orthorhombic) were used as starting materials. Mechanical activation was carried out by using a multi-ring type mill.

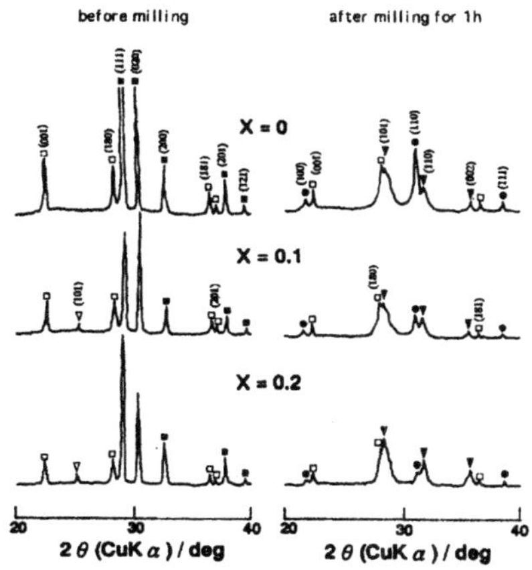


*Figure 7.7.* X-ray patterns of the precursor of PMN-PT.
()PbO(massicot); ()) PbO(litharge); (□)$Nb_2O_5$; (Δ)$TiO_2$(anatase); ()perovskite.

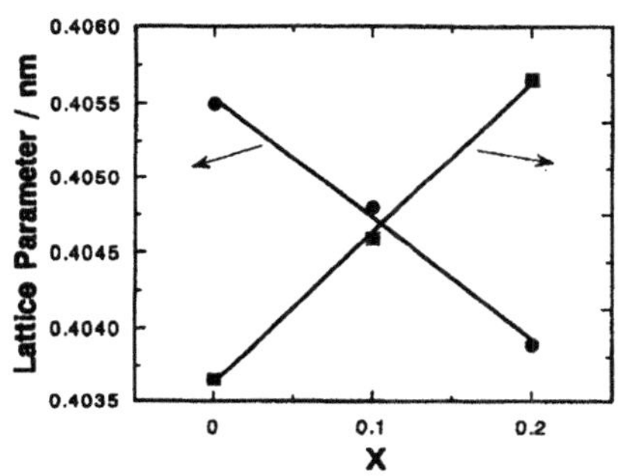


*Figure 7.8.* Lattice parameter and average grain size of the precursor of PMN-PT after calcined at 850°C for 4 h.

The specific surface area is insignificantly changed by milling, i.e., at most with an increase by 35%. Oxygen 1s XPS spectra reveal that the multimodal peaks around 531 eV for the mixture before milling turn into a quasi-single peak at 530.3 eV, being attributed to the Pb-O-multicomponent B-site ions of the perovskite $ABO_3$ (Fig. 7.6). Together with the XRD patterns, this demonstrates that new chemical bonds leading to a perovskite phase are already formed during milling.

Although the formation of columbite, $MgNb_2O_6$, was not detected by X-ray diffractometry, low reactive MgO was very likely to have been combined with $Nb_2O_5$ to give a precursor, thus, facilitating the further reaction to the final form, PMN–PT [22]. In a soft mechanochemical process for PMN–PT, separate preparation of columbite is, therefore, not necessary.

As shown in Fig. 7.7, a pure perovskite phase was obtained after calcining the milled mixture at 850°C for 4 h in air. On the other hand, a cubic pyrochlore ($P_3N_4$) phase was predominant with the coexistence of perovskite and PbO (litharge) on the calcined products from non-milled mixture. When MgO was used instead of $Mg(OH)_2$, no significant change of

O1s XPS peak was observed. This shows the preference of $Mg(OH)_2$ to MgO during milling for the mechanochemical reaction toward a complex compound. The lattice constant of the cubic perovskite phase decreases linearly with x in the calcined product, as shown in Fig. 7.8. This is an indication of the formation of uniform PMN–PT solid solution.

Fig. 7.9 shows the temperature dependence of the dielectric constant and dielectric loss at 1 kHz for the PMN–PT ceramics obtained by sintering the calcined powders from a soft-mechanochemical route at 1200°C for 2 h. A diffuse phase transition, being typical for a relaxor, is observed for each ceramics. As x increases from 0 to 0.2, the maximum dielectric constant, $K_{max}$, increases from 13000 to 27000. The temperature correspondent to $K_{max}$,

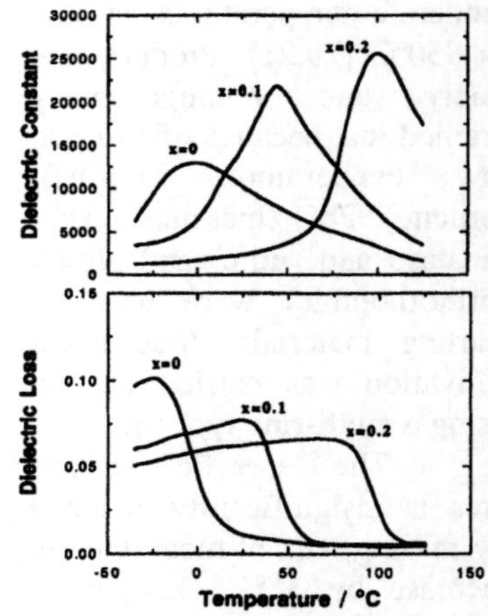


*Figure 7.9.* Temperature dependence of dielectric constant and loss of the samples fired at 1200°C for 2 h.

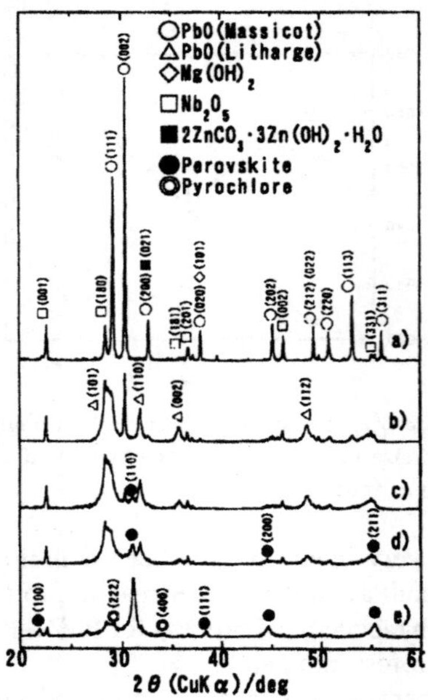


*Figure 7.10.* X-ray patterns of intact and as-milled mixtures for 0.3PMN-0.7PZN. Milling time was (a)0, (b)5, (c)30, (d)60 and (e)180 min.

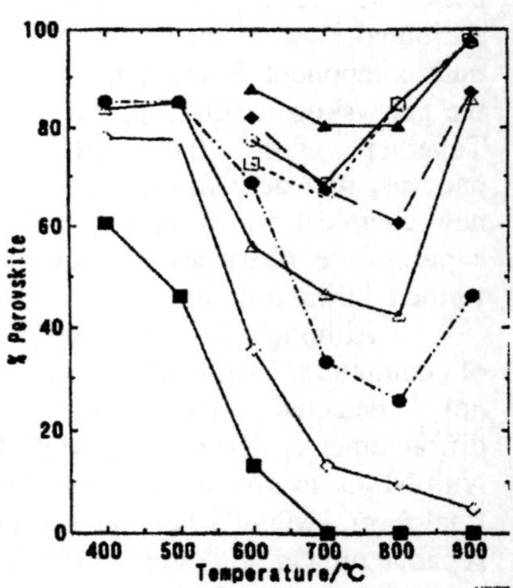


*Figure 7.11.* Plots of percent perovskite against firing temperature for mixtures milled for 180 min(x is ()0, (□)0.2, ()0.4, (♦)0.6, (Δ)0.7, (•)0.8, () 0.9, ()1.0).

i.e., $T_{max}$, also increases with x from -1°C to 99°C.

The principle was further extended to other perovskites, which are even more difficult to synthesize, such as $(1-x)Pb(Mg_{1/3}Nb_{2/3})O_3 \cdot xPb(Zn_{1/3}Nb_{2/3})O_3$ (PMZN), where a fraction $x$ of Mg is replaced by Zn [23,24]. As shown in Fig. 7.10, we obtained phase pure perovskite up to x=0.7 in one step from a stoichiometric mixture comprising those components used for PMN and basic zinc carbonate. During heating, perovskite formed as a result of milling, disappeared first, but emerged again by further heating, as shown in Fig. 7.11, due to the migration of the stabilizing component, Mg, during heating.

Thus, soft mechanochemical processing is versatile for the fabrication of complex oxides such as phase pure PMN-PT. This is mainly attributed to the formation of principal short range ordering, Mg–O–Nb, necessary for the final product at the state of precursor during the mechanical treatment of the starting mixture.

*Table 7.1.* Sample sols.

| Samples | Ca/P of the starting materials | Concentration of $NH_3$, (M) |
|---|---|---|
| CP1 | 1.00 | 0 |
| CP1.25 | 1.25 | 0 |
| CP1.37 | 1.37 | 0 |
| CP1.5 | 1.5 | 0 |
| CP1.67 | 1.67 | 0 |
| CP2 | 2.0 | 0 |
| CP1.37a1 | 1.37 | 0.004 |
| CP1.37a2 | 1.37 | 0.04 |
| BS-tris[a] | 1.00 | 0 |

[a] Buffer solution milled in tris.

## 7.4. Hydroxyapatite

Hydroxyapatite (HAp) and its isomorphous modifications are important and prospective materials in bioceramics. There are some mechanochemical studies on HAp. In [25] HAp was prepared from calcium hydrated orthophosphates and calcium hydroxide or oxide by mechanical activation of mixtures in planetary mill. The authors

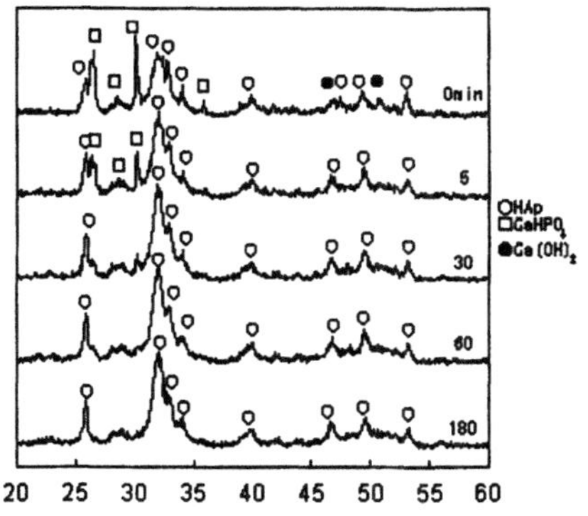


*Figure 7.12.* X-ray pattern of the product CP1.67 after milling for different time.

[26] succeeded in obtaining HAp starting from brushite and calcium carbonate. They also doped HAp by $Sr^{2+}$ [27]. However, crystallinity of mechanochemically prepared products was low unless the products were heated at 500÷1000°C.

Here, a new method for the preparation of an aqueous, well-dispersed sol of HAp nano particles directly via a mechanochemical route is described. It should be noted, that such products are very suitable for usual ceramic purposes as well.

200 $cm^3$ of 1.05 M aqueous solution of $H_3PO_4$ was mixed with varying amount of $Ca(OH)_2$ powder. The Ca/P molar ratio of the starting materials was between 1 and 2. The sol of the stoichiometric mixture with Ca/P=1.67 was mixed with varying amount of a dispersant, ammonium polycarboxylate (Dispex A40, Allied Colloid), to give a sol of 26.1 wt. % of solid after completion of the reaction. The mixture was subsequently subjected to milling by a multi-ring ultrafine grinding mill (Micros-0, Nara Machinery) at 1250 rpm for 3 h. Detailed description of the mill is given elsewhere [28]. Samples are listed in Table 7.1.

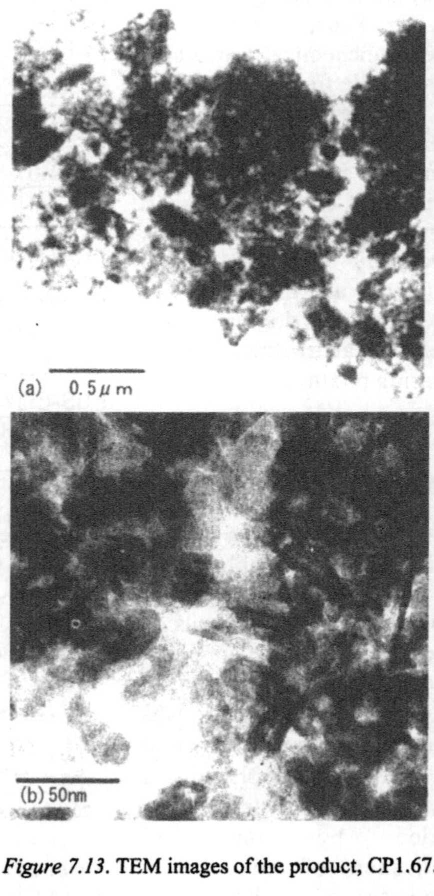


*Figure 7.13.* TEM images of the product, CP1.67.

As shown in Fig. 7.12, $CaHPO_4$ and $Ca(OH)_2$ are present in intact mixtures. After milling for 3 h all the X-ray diffraction peaks belong to hydroxyapatite, without any second phases, such as monetite, ($CaHPO_4$), brushite, ($CaHPO_4.2H_2O$), or the reactant, $Ca(OH)_2$. Crystallite size, obtained by Wilson's method, was equal to 17 nm without appreciable lattice strain [29].

As shown in Fig. 7.13*a*, most of the observed particles are equiaxed. Their average diameter determined from electron micrographs is of 80 ± 140 nm. A higher magnification reveals, however, that particles described above are still agglomerates of much smaller primary particles, 4.0 ± 9.5 nm, as shown in Fig. 7.13*b*. The value is close to the crystallite size estimated from the line broadening of the X-ray diffraction peaks, given above. Therefore, this method gives rise to HAp crystallites with their average size below 20 nm. These nanocrystals are well dispersed in the form of agglomerates of 100 nm size.

The observed molar ratio, Ca/P, was 1.59±0.01 for CP1.67 sample. Since the Ca/P ratio was always smaller than the stoichiometry of HAp (1.67) the product as prepared can be regarded as a slightly Ca deficient HAp.

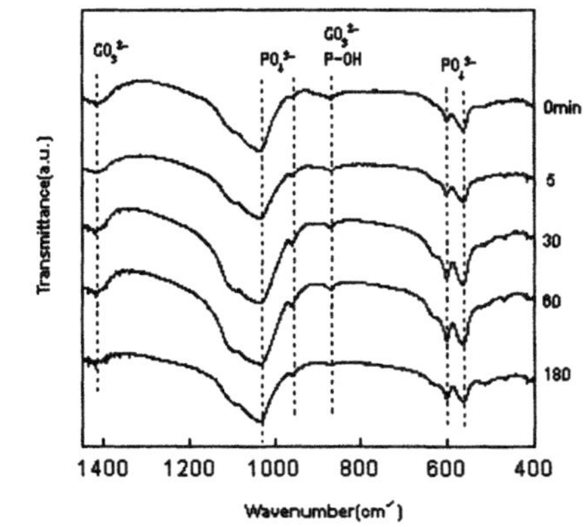


*Figure 7.14.* IR spectra of the product CP1.67 after milling for different time.

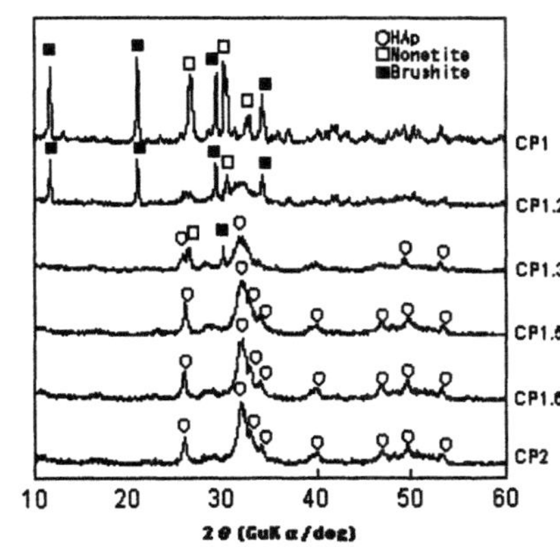


*Figure 7.15.* X-ray patterns of the products CP1-2.

On the FTIR profiles of as-milled sol (Fig. 7.14), absorption bands at 565, 602, 965 and 1037 $cm^{-1}$ are observed. All these bands belong to $PO_4^{3-}$ of HAp. Corresponding bands of the second phases appear at appreciably different wavenumber, e.g., at 1070 $cm^{-1}$ for monetite, instead of 1037 $cm^{-1}$ for HAp [30]. A small band at 875 $cm^{-1}$ is also observed, which is attributed to $CO_3^{2-}$ and $HPO_4^{-2}$ [27-31]. It is well known that addition of $Ca(OH)_2$ into $H_3PO_4$ results in monetite and, subsequently, in brushite [32,33]. The final stage toward HAp formation from brushite usually requires heat treatment. In the case of $Ca(OH)_2$ milled in $H_3PO_4$·aq, however, a proton from $HPO_4^{2-}$ ion can transfer to an $OH^-$ ion of $Ca(OH)_2$ to induce dehydration, like in many other mechano-chemical acid-base reactions [34-36].

There are several possibilities to explain existence of $HPO_4^{2-}$ band in FTIR spectra. One of them is due to incomplete deprotonation of $HPO_4^{2-}$ to

$PO_4^{3-}$, $HPO_4^{2-}$-containing phases, like monetite or brushite, may coexist with each other, but the presence of second phases is rather improbable since the corresponding IR band is not registered in the IR atlas [30] as well as no X-ray diffraction peaks of these phases was observed. Another possible source of $HPO_4^{2-}$ is Ca deficiency, which was actually observed, e.g., Ca/P=1.59±0.01 for CP1.67 sample. Contribution of surface OH groups bound to phosphor, i.e., P-OH, to the $HPO_4^{2-}$ band is not excluded, similar to many other surface OH groups on metal oxides. A decrease in the peak intensity after annealing the sample at 900°C for 3 h was actually observed.

As shown in Fig. 7.15, all the X-ray diffraction peaks of CP1.5, CP1.67 and CP2 samples belong to HAp. As the Ca/P molar ratio decreases from 1.375, the peak intensity of HAp decreases appreciably, while that of monetite and brushite increases. The effect of $NH_3$ adding is shown in Fig. 7.16. The peak intensity of monetite and brushite decreases or even disappears by the addition of $NH_3$. According to Table 7.2, the final pH of sol was changed from acid to base by adding $NH_3$.

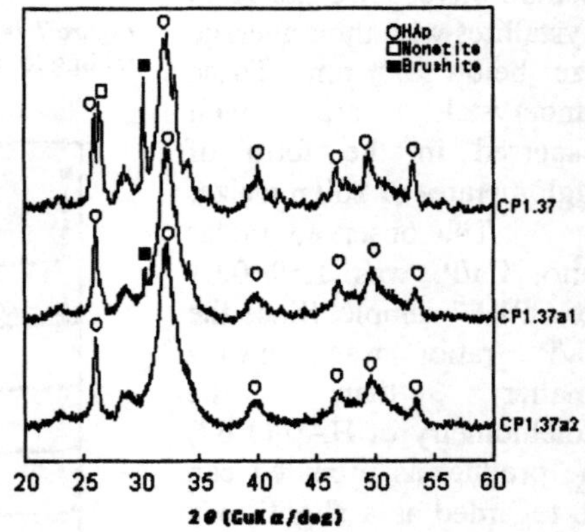


*Figure 7.16*. X-ray patterns of the products CP1.37, CP1.37a1 and CP1.37a2 samples.

The X-ray patterns of milling brushite in tris buffer solution are shown in Fig. 7.17. For 1 h stirring, the profile does not change. However, after milling for 3 h, only diffraction peaks of Hap was observed. Phase pure HAp was not obtained even by ageing for 24 h. The pH of the sol changed from 8.28 to 7.62 during the reaction. This fact indicates that pH value plays a significant role on crystalline state of the final products.

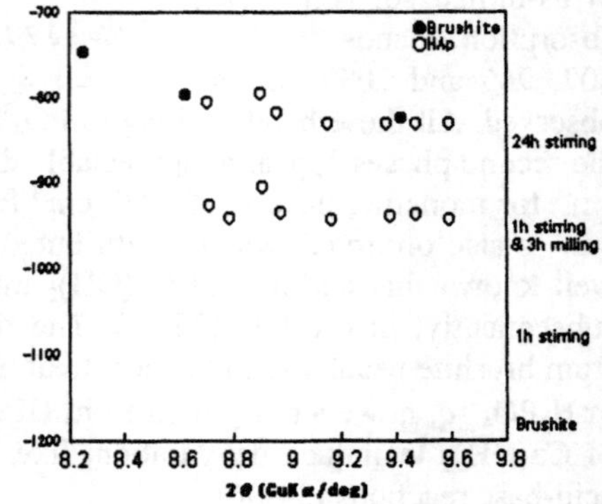


*Figure 7.17*. X-ray patterns of the product BS-tris.

Merit of the present 'in situ' preparation of HAp via a wet mechanochemical route is at least two-fold, i.e, (1) to assist deprotonation from brushite to form hydroxyapatite, and (2) to keep high dispersion and low viscosity of sol. The descibed method of nano-sol preparation will open a new way of applying HAp for casting, injection or to prepare thin films by a simple dip coating procedure.

*Table 7.2.* pH and final Ca/P of CP1-2 samples.

| Samples | Final pH of sol | Final Ca/P |
|---|---|---|
| CP1 | 5.95 | — |
| CP1.25 | 5.75 | — |
| CP1.37 | 5.90 | 1.38±0.02 |
| CP1.5 | 6.70 | 1.59±0.03 |
| CP1.67 | 9.34 | 1.59±0.01 |
| CP2 | 12.29 | 2.02±0.03 |
| CP1.37a1 | 7.25 | 1.46±0.02 |
| CP1.37a2 | 10.53 | 1.44±0.05 |

The detail studies of mechanochemical interaction in the course of activation of $Ca(H_2PO_4)_2{\cdot}H_2O$+CaO mixtures showed that the degree of interaction is dependent on the amount of energy supplied to the reaction zone [37]. If the energy is less than 15 kJ/g, the phases of an aqueous phosphate-calcium system are formed (the lower part of the diagram on Fig. 7.18). As the amount of energy increases, dehydration of the phases starts in system: $CaHPO_4$ is transformed into pyrophosphate $Ca_2P_2O_7$, while the activation of $Ca(H_2PO_4)_2{\cdot}H_2O$ even for 60 min doesn't lead to any changes.

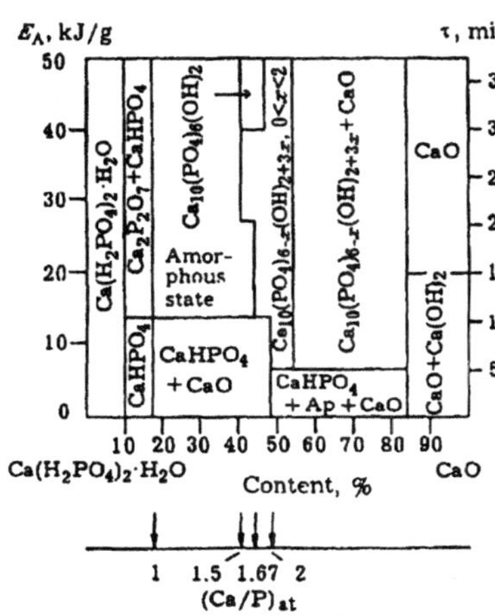


*Figure 7.18.* Kinetic energy diagram of $Ca(H_2PO_4)_2{\cdot}H_2O$-CaO system [37].

The stoichiometric hydroxylapatite $Ca_{10}(PO_4)_6(OH)_2$ was formed after 30 min activation (40 kJ/g) of the mixture with Ca/P=1.67. The phosphates remains amorphous even in the case of the mixture with Ca/P=1.5 activated for 1 h. This can be explained as a

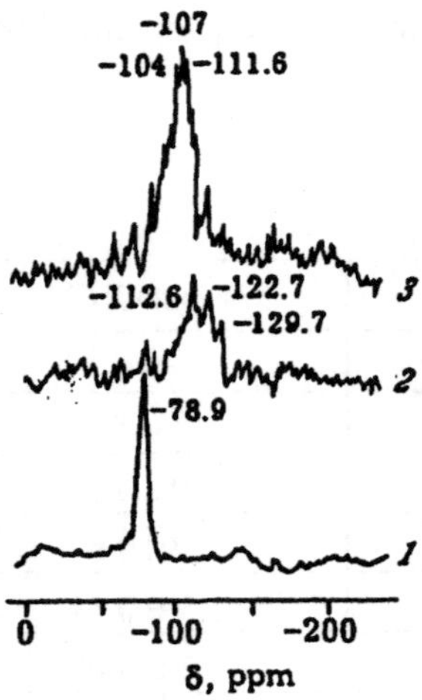


*Figure 7.19.* $^{29}Si$ MAS NMR spectra:
1 – initial mixture, $3Ca(H_2PO_4)_2 \cdot H_2O$+7CaO+$Si_{el}$,
2 – after its activation for 30 min,
3 – sample 2 after annealing at 900°C.

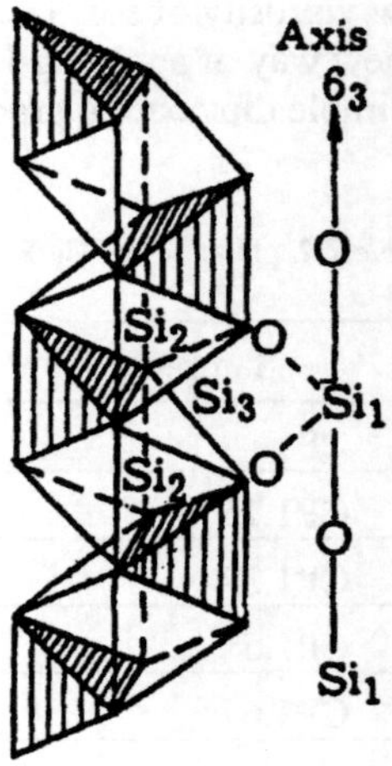


*Figure 7.20.* The scheme of possible localization of silicon in the apatite structure.

result of stabilization of amorphous phase by water. The polarization of OH bond of water in vicinity of phosphate ion causes an increase of acid properties of water and thus strengthens hydrogen bonds with anion which hinders the dehydration of the product [38].

The convenient method of preparation of substituted forms of HAp is described in [39,40]. The elementary silicon was added to $3Ca(H_2PO_4)_2$+CaO mixture which was then mechanically activated in planetary mill for 25 min. The sample was studied by $^{29}Si$ MAS NMR spectroscopy. It was shown that three bands of silicon are presented in spectra with chemical shifts of -129.7, -122.7, -112.6 ppm (Fig. 7.19).

According to [41], the chemical shift value (δ) and bond angle (θ) for the T-O-T bond (where T= Si, Al and others) are connected by the equation: $\delta = -25.44 - 0.57\ \theta$. According to this equation, chemical shifts in the $^{29}Si$ MAS NMR spectra of the synthesized apatite correspond to Si-O-T bridges with the angles 180°, 170° and 150°, respectively. It can be assumed that the NMR signal at –129.7 ppm is determined by the localization of silicon cation on the $6_3$ axis of HAp in Si-O-Si or Si-O-Si-OH bridges with the angle of 180°. Two other silicon ions occupy the vacant places between the phosphate tetrahedrons and calcium ions with the bond angles between oxygen bridges ~170° and 150°, respectively. After annealing these samples at 900°C, apatite transforms into $\beta$-$Ca_3(PO_4)_2$, and the $^{29}Si$ MAS NMR spectra are changed (Fig. 7.20).

The hydroxilapatite with silicon ions in the structure are characterized by higher compression strength, compared to nondoped hydroxilapatite; thus, it can be used as a material for bioceramics.

## 7.3. Heteropolyacids

Iso- and heteropolyacids represent an important class of inorganic compounds. They form a separate structural class of complexes based on quasi-octahedrally coordinated d-element atoms of the V and VI periods of Mendeleev's Periodic Table [42].

The first attempts to represent the composition of heteropolyanions were based on Verner's coordination theory according to which heteropolyacids are considered to be the complexes containing octahedrally coordinated heteroatom bound to $MO_4^{2-}$ or $M_2O_7^{2-}$ anions as ligands, for example $H_8[Si(W_2O_7)_6]$. This was confirmed by the synthesis of the cesium salt of this acid, $Cs_8[Si(W_2O_7)_6]$. However, Pauling demonstrated later that the structure of this heteropolyacid should be considered as consisting of the central tetrahedron $XO_4$ surrounded by 12 $WO_6$ octahedrons [43]. He proposed to use the formula $H_4[SiO_4W_{12}O_{18}(OH)_{36}]$, which agrees with the observed basicity of this acid. However, Pauling considered $MO_6$ octahedrons to be bound only by their vertices.

Keggin [44] showed that octahedrons could be linked with the tetrahedrons $XO_4$ both by the vertices and by edges also that was later confirmed by X-ray crystallography data. These structures were called Keggin structures.

According to the modern notions, heteropolyacids (polyoxoanions) are synthesized by the condensation of several simple anions with water evolving. These negatively charged particles are composed either of octahedrons or tetrahedrons, or both of them. Octahedrons and tetrahedrons are formed by six or four atoms, respectively, around central atoms. Octahedrons and tetrahedrons are linked to each other by edges or vertices. An atom of metal or non-metal can serve as the central atom.

The most common method to synthesize polyoxoanions is to acidify monomeric oxoanions and the compounds containing the required heteroatoms, for example:

$$7MoO_4^{2-} + 8H^+ \rightarrow [Mo_7O_{24}]^{6-} + 4H_2O$$
$$6MoO_4^{2-} + Cr(H_2O)_6^{3+} + 6H^+ \rightarrow [Cr(OH)_6Mo_6O_{24}]^{3-} + 6H_2O$$
$$12WO_4^{2-} + HPO_4^{2-} + 23H^+ \rightarrow [PW_{12}O_{40}]^{3-} + 12H_2O, \text{ etc.}$$

Chemical equilibrium constants and formation rate constants are rather large, so polyanions can be isolated as salts from the acidified stoichiometric mixtures of components at room temperature.

Similarly to the above described methods of the preparation of jointly precipitated hydroxides (see Chapter 2), important role is played by the order in which the reagents are added to the reaction mixture. If the order is changed, the products with different structures are obtained. For example,

$SiO_3^{2-}$, $WO_4^{2-}$, then $H^+ \rightarrow \alpha$ - $[SiW_{12}O_{40}]^{4-}$
$WO_4^{2-}$, $H^+$, then $SiO_3^{2-}$, $H^+ \rightarrow \beta$ - $[SiW_{12}O_{40}]^{4-}$.

The mixture is usually acidified by adding mineral acids; polyanion can be isolated by adding the corresponding cation, usually an alkaline metal cation or ammonium, with the formation of salt.

At present, more than 100 heteropolyacids of different structure and composition are known [45]. They are very strong proton acids; they are stronger than perchloric, sulphuric and hydrochloric acids. At the same time, they possess rather high oxidative potential and can cause redox transformations. Many heteropolyacids are very stable from chemical and thermal viewpoints; they are well soluble in water and oxygen-containing solvents. Thanks to their properties, they are applied in many areas, but mainly in analytical chemistry and catalysis. The most important are heteropolyacids with Keggin type structure of the general formula $H_nXM_{12}O_{40}$, where X is heteroatom, M=Mo, W, V or their mixture. Heteropolyacids with Dawson structure of the general formula $H_nX_2M_{12}O_{62}$ are of less importance.

Heteropolyacids, such as $H_3PW_{12}O_{40}$, $H_3PMo_{12}O_{40}$, $H_4SiW_{12}O_{40}$, $H_4SiMo_{12}O_{40}$, and $H_6P_2W_{18}O_{62}$ in the form of ammonium salt, are produced in industry. However, the synthesis methods insufficiently fit for modern technological, economical and ecological requirements. Therefore, new methods are being developed.

One of the methods to synthesize heteropolyacids is the direct interaction of the oxides of ligand atoms (tungsten, molybdenum, vanadium) with oxides or acids of heteroatoms (phosphorus, silicon, etc.). However, because of low reactivity of oxides, these processes are slow and energy-consuming (heating for tens hours is required); quantitative yield is rarely achieved.

Mechanical activation applied to the mentioned reaction was found to be very efficient [45]. For example, molybdenum oxide mechanically activated for several seconds is readily dissolved in hot solution of phosphoric acid to give heteropolyacid. By varying the interaction conditions, one succeeds in obtaining heteropolyacids with different molybdenum to phosphorus ratios. $H_3PMo_{12}O_{40}$ can be synthesized in the individual pure state, and $H_6P_2Mo_{18}O_{62}$ with the purity of 94-98 %. Mo – P mixtures, which are most easily prepared, can be used as acid catalysts without separating into components. Besides the simplicity of preparation, another advantage of this method, consists in the possibility of recovery of worked-out catalysts.

Even more substantial effect of mechanical activation on the reactivity is exhibited by the mixtures of $MoO_3$ and $V_2O_5$. After mechanical activation, these mixtures are rapidly dissolved in $H_3PO_4$ solutions giving $H_{3+n}PMo_{12-n}V_nO_{40}$ (n = 1÷6). Moreover, some acids can be obtained by direct activation of the mixtures of molybdenum oxide, vanadium oxide and phosphoric acid. This technology allows to exclude the stage of heteropolyacid isolation from aqueous solution by partial evaporation, that makes it very productive and energy-saving. The salts of heteropolyacids can be prepared by this procedure, too.

The reactivity of tungsten oxide is also slightly increased after it mechanical activation. In the case of activated $WO_3$, 10 % of oxide reacts with the solution of phosphoric acid, while in the case of activated $WO_3$ - $V_2O_5$ mixtures, the yield increases up to 13 %.

The yield of silicon-containing heteropolyacids is lower than that of phosphorus-containing one; it depends on the amount of silica in the mixture under activation, that one can see in Table 7.1 given below:

*Table 7.3.* Yield of heteropolyacids after mechanical activation of oxide mixtures.

| Mixture* | Mixture composition, % | HPA yield, % |
|---|---|---|
| $MO_3 - SiO_2$ | 92.3 – 7.7 | 37 |
| | 82.8 – 17.2 | 60 |
| | 67 – 33 | 79 |
| $WO_3 - SiO_2$ | 91.9 – 8.1 | 7.7 |
| | 83 – 17 | 9 |
| | 71 – 29 | 13 |
| $WO_3 - V_2O_5 - SiO_2$ | 77 – 6 – 17 | 20 |
| | 65.8 – 5.2 – 29 | 20 |

***After mechanical activation, mixtures were treated with boiling water for10 min.**

Among other heteropolyacids, $H_9AlMo_6O_{24}$ and $H_9FeMo_6O_{24}$ were obtained according to similar procedure in solutions. These acids do not exist in their individual form, they were isolated as tetrabutylammonium salts.

The above-described method of heteropolyacid synthesis is distinguished by high productivity, complete use of expensive initial reagents, the absence of any wastes, low energy consumption (about 0.4 kW/h) [46].

## *References*

1. Godovikov, Alexandr, *Mineralogy*. Moskva: Nedra, 1975.
2. Devyatkina E.T., Avvakumov E.G., Kosova N.V., Lyakhov N.Z. Mechanical activation in the synthesis of cordierite. Neorganicheskie materialy 1994; 30: 237-40.
3. Prokofyev V.Yu., Kunin A.V., Ilyin A.P. et al. Application of mechanochemical methods to the synthesis of cordierite catalyts supports. Zhurn. Prikl. Khimii 1997; 79: 1655-60.
4. Prokofyev V.Yu., Ilyin A.P., Kunin A.V. et al. Mechanochemical synthesis of cordierite from natural and synthetic raw materials. Chem. Sustainable Development 1998; 6: 137-70.
5. Atanovska M., Djuricic M., Milosevic S. et al. Kinetics and mechanism of mechanochemical activations of the $2MgO–2Al_2O_3–5SiO_2$ system. Science of Sintering 1992; 24: 185-91.
6. Awano M., Takagi H., Kuwahara Y. Grinding effects on the synthesis and sintering of cordierite. J. Am. Ceram. Soc. 1992; 75: 2535-40.
7. Filio J.M., Sugiyama K., Kassi N. et. al. Effect of dry mixed grinding of talk, kaolinite and gibbsite on preparation of cordierite ceramics. J. Chem. Ing. Japan 1994; 26: 565-69.
8. Avvakumov E.G., Devyatkina E.T., Kosova N.V. et al. Novel mechanochemical method of the preparation of cordierite and support based on it. Kinetika i Kataliz 1998; 39: 722-25.
9. Avvakumov E.G., Gusev A.A., *Cordierite as a Prospective Ceramic Material.* Novosibirsk: SO RAN, 1999.
10. Avvalumov E.G., Devyatkina E.T., Kosova N.V. Mechanochemical reactions of hydrated oxides. Fiziko-tekhnicheskie Problemi Pererabotki Poleznikh Iskopaemikh 1993; 2: 541-58.
11. Avvakumov E.G. Soft mechanochemical synthesis as the basis for new chemical technologies. Chem. Sustainable Development. 1994; 2: 541-58.
12. Ovcharenko N.A., Kornilovich B.Yu., Ovcharenko F.D. Investigation of the kinetics of hydrothermal synthesis of monoclinic celsian. Dokl. AN USSR: B 1978; 1: 37-39.
13. Kornilovich B., *Structure and Surface Properties of Mechanochemically Activated Silicates and Carbonates*. Kiev: Naukova Dumka, 1994.
14. Kornilovich B.Yu., Zapolsky A.K., Lysenko V.I., Malyarenko V.V. The effect of mechanochemical activation on solid-phase synthesis of barium aluminosilicates. Izv. AN SSSR. Neorg. Materialy 1988; 24: 984-87.
15. Vishnev A.A., Klimov E.G., Kolbanev I.V. et al. Specific features of the synthesis and the properties of high-temperature superconductor $YBa_2Cu_3O_x$ obtained by mechanochemical activation of the initial oxides. Superconductivity: Physics, Chemistry, Technology 1990; 3: 2390-99.
16. Hainovsky N.G., Pavlyukhin Yu.T., Boldyrev V.V. Application of mechanical activation to the synthesis of high temperature superconductors. Mat. Science Eng. 1991; B8: 283-86.
17. Sobolev A.S., Kovalenko V.M., Fotiev A.A., Kosheeva S.N. The formation of $YBa_2Cu_3O_{7-\delta}$ in thermal treatment of mechanically activated stoichiometric mixture. In *Physico-chemical Foundations of the Synthesis and Properties of High-Temperature Superconducting Materials*. Sverdlovsk: UrO AN SSSR, 1990; 153-55.
18. Panova T.I., Malysheva S.I., Savchenko E.P., Glushkova V.B. Synthesis of complex cuprates in the system La-Ba-Cu-O with mechanical activation of the components. Neorg. Materialy 1999; 35: 1487-90.
19. Rozental O.M., Fotiev V.A., Avvakumov E.G., Fedanchin E.I. Mechanosynthesis of yttrium and barium cuprates. Proc. All-Union Scientific and Technological Conference "Mechanochemical synthesis", Vladivostok, 1990; 112-14.

20. Baek J.G., Isobe T., Senna M. A new fabrication technique for perovskite $0.9Pb(Mg_{1/3}Nb_{2/3})$-$0.1PbTiO_3$ ceramics via a soft-mechanochemical route. J. Am. Ceram. 1997; 80: 973-81.
21. Baek J.G., Gomi K., Isobe T., Senna M. Better ferroelectric ceramics starting from a soft-mechanochemical route. Chem. Sustainable Development 998; 6: 171-74.
22. Shinohara S., Baek J.G., Isobe T., Senna M. Synthesis of phase pure $Pb(ZnxMg_{1-x})_{1/3}Nb_{2/3}O_3$ up to x=0.7 from a single mixture via a soft-mechanochemical route. J. Am. Ceram. Soc. 2000; 83: 2308-10.
23. Hamada K., Isobe T., Senna M. Comparative studies on the mechanochemical synthesis of $MgTiO_3$ precursors by milling various mixtures containing oxides and hydroxides. J. Mater. Sci. Lett. 1996; 15: 603-5.
24. Baek J.G., Isobe T., Senna M. Mechanochemical effects on the precursor formation and microwave dialectic characteristics of $MgTiO_3$. Solid State Ionics 1996; 90: 269-27.
25. Chaikina M.V., Shapkin V.L., Kolosov A.S., Boldyrev V.V. Mechanochemical synthesis fluor- and chlorapatite. Izv.SO AN SSSR, ser.khim.nauk 1978; 3: 96-101.
26. Toriyama M., Ravaglioli A., Krajewski A., Galasse C., Roncari E, Piancastelli A. Slip casting of mechanochemically synthesized hydroxyapatite. J. Mater. Sci. 1995; 30: 3216-21.
27. Yokogawa Y., Toriyama M., Kawamoto Y., Suzuki T., Nishizawa K., Nagata F., Mucalo R. Preparation of calcium-strontium apatite through mechanochemical method. Chem. Lett. 1996; 1: 91-2.
28. Hamada K., Yamamoto S., Nagao M., Senna M. Measurement of compressive and shear forces in multi-ring media mill: J. Chem. Eng. Jpn. 1997; 30: 756-59.
29. Peiser H.S., Rooksby H.P., Wilson A.J.C., *X-Ray Diffraction by Polycrystalline Materials*. London: Institute of Physics, 1955.
30. *The Aldrich Library of FTIR Spectra,* 1. Ed., Vol. 2, ed. Pouchert C.J., Milwaukee: Aldrich Chem. Comp., 1985.
31. Vallet-Regi M., Rodriguez-Lorenzo L.M., Salinas A.J. Synthesis and characterization of calcium deficient apatite. Solid State Ionics 1997; 101/103: 1279-85.
32. Lazic S., Katanic-Popovic J., Zec S., Miljevic N. Properties of hydroxyapatite crystallized from high temperature alkaline solutions. J. Cryst. Growth 1995; 147: 124-28.
33. Ijima M., Kamemizu H., Wakamatsu N., Goto T. J. Transition of octacalcium phosphate to hydroxyapatite in solution at pH 7.4 and 37°C. Cryst. Growth 1997; 181: 70-8.
34. Watanabe T., Isobe T., Senna M. Mechanisms of incipient chemical reaction between $Ca(OH)_2$ and $SiO_2$ under moderate mechanical stressing, III. Changes in the short range ordering through the mechanical and thermal processes. J. Solid State Chem. 1997; 130: 284-89.
35. Senna M. Mild mechanical stressing on mixtures to a precursor for homogeneous complex oxides. Chemistry Reviews 1998; 23: 263-84.
36. Fujiwara Y., Isobe T., Senna M., Tanaka J. Effects of reduced coordination number for Ca on the electron redistribution during Ca-O-Si bridge bonding from CaO or $Ca(OH)_2$ and $SiO_2$. J. Phys. Chem. 1999; 103: 9842-46.
37. Chaikina M.V. Mechanochemical synthesis of apatites and orthophosphates. In *Mechanochemical Synthesis in Inorganic Chemistry,* ed. Avvakumov E.G., Novosibirsk: Nauka, 1991.
38. Chaikina M.V. Mechanochemical synthesis of phosphates and apatites – a new way of preparation of complex materials. Phosphorus Research Bulletin 1997; 7: 35-8.
39. Chaikina M.V. The features of chemical interaction in multicomponent system during mechanochemical synthesis of phosphates and apatites. Chem. Sustainable Development 1998; 6: 135-44.
40. Makatun V.N., *Chemistry of Inorganic Hydrates*. Minsk: Nauka i Tekhnica, 1985.
41. Klinowski J. Nuclear magnetic resonance studies of zeolites. Progress in NMR spectroscopy 1984; 16: 237-309.

42. Pope, Michael Thor, *Hetepoly and Isopoly Oxometalates.* Berlin, Heidelberg: Springer-Verlag, 1983.
43. Pauling L. C. The principles determining the structure of complex ionic crystals. J. Am. Chem. Soc. 1929; 51: 2868-75.
44. Keggin J.F. Structure and formula of 12-phosphotungstic acid. Proc. Roy. Soc. 1934; A144: 75-100.
45. Maksimov G.M. Achievements in the synthesis of polyoxometallates and investigation of heteropolyacids. Uspekhi Khimii 1995; 64: 480-92.
46. Maksimov G. M., Molchanov V. V., Gaidin V.V. New technologies of heteropolyacids and catalysts based on them. Khim. Promischlennost 1998; 10: 3-5.

# Chapter 8

# SOME FEATURES AND POSSIBLE MECHANISMS OF MECHANOCHEMICAL REACTIONS OF HYDRATED OXIDES

On the basis of experimental data for various reactions described above, one can outline several main features and regularities of soft mechanochemical reactions and propose their possible mechanisms.

First of all, it should be noted that the regime of these reactions is not uniform within reaction time. As the initial hydrated oxides are solids, the reaction starts via the solid phase mechanism. The solid phase mechanism is confirmed by the observed influence of initial disperse state on the interaction rate, the participation of surface acid-base centers in the reactions, the existence of concentration gradient in the surface layers of the reagents, etc. [1-8].

However, the formation of water as a reaction product changes the regime of interaction between the components, providing the possibility to interact via the dissolved state. In this case, the solubility of initial and final products in water become essential. Therefore, it is possible to consider the mechanochemical reactions from the viewpoint of the theory of reactions in water solutions. Three types of reactions are discussed usually: a) acid-base reactions, b) redox reactions, and c) exchange and substitution reactions. There are no clear differences between the mentioned reactions; all of them can be attributed to some extent to acid-base interactions. As the important parameter characterizing redox reactions is the redox potential of the system, its change leads to the changes of acid-base properties of the system. This allows to consider the reactions which take place in this system from the viewpoint of acid-base mechanism. Exchange or substitution reactions in complex ions are much simpler for interpretation than redox reactions. These reactions involve the substitution of one ligand by another; but in more general aspect they can be considered as acid-base reactions between metal ions and Lewis acids. All these aspects are presented in the considerations described below.

## 8.1. Mechanisms of formation of new bonds

A bond energy between cation and oxygen ion in oxide is a measure to determine whether an $OH^-$ group on solid surface is an acid or a base. Therefore, the acid-base reaction is always possible to occur when two different Me-O(H) bonds are brought into contact (Fig. 8.1).

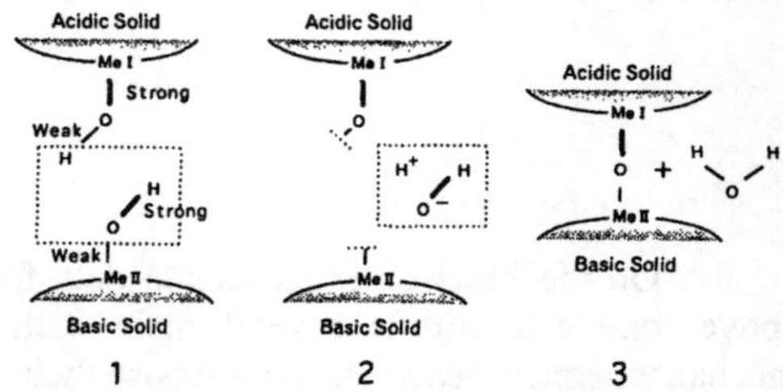


*Figure 8.1.* A schematic illustration of mechanical dehydration preceding HMB formation.

Hydroxyl groups on the solid surface become more polarized when they are combined with atoms at their low-coordinate state. This is generally understandable because of displacing of the electron density. The low-coordination states are metastable and highly reactive. They are formed during milling of a mixture of hydrated oxides at the contact points, where plastic deformation can preferably take place. In spite of that the hydrated oxides are the brittle material, they are plastified when their particles are small. The occurrence of plastic deformation of brittle substances has been thoroughly studies by Hess [9]. On the basis of change of stress-strain curve of compressed single particles with different sizes, he showed the general tendency of the particles to be plastified, when their size becomes smaller than a certain critical value. Plastic deformation itself cannot bring about a direct chemical reaction, but it causes mechanochemical reaction through the increase in the number of low-coordinate atoms.

The appearance of low-coordinate atoms is the reason of the increasing of the basic centers during the mechanical activation of hydroxides. This suggestion is confirmed by the study of basic centers on $Ca(OH)_2$ surface [4,6]. The basic Lewis centres ($O^{2-}$) were obtained on the $Ca(OH)_2$ surface by using sorption of gas $NH_3$ and $CO_2$ [5] (Fig. 8.2).The concentration of basic centres increases up to a certain value and then remains constant under mechanical activation of single $Ca(OH)_2$, while it grows more significantly and has the maximum for $Ca(OH)_2$+$SiO_2$·aq activated mixtures. It means that an increase in the basicity at the initial stage of activation of the mixture is connected not only with the formation of new surface but also with the generation of new centres at the contact points, where plastic deformation occurs.

The reaction in this mixture was also studied by IR spectroscopy. The formation of the product, calcium silicate, is characterized by the appearance of a new intense band at 1005 $cm^{-1}$. A comparison of data on basicity and the IR spectra shows that the silicate formation is accompanied

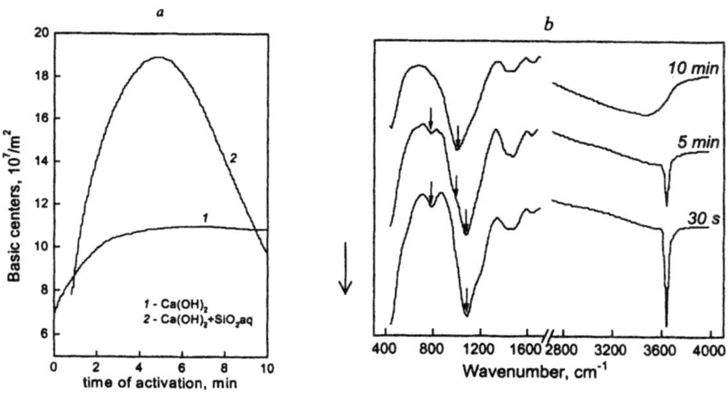


*Figure 8.2.* The changes of basicity (*a*) and IR spectra (*b*) for the $Ca(OH)_2$+$SiO_2$aq mixture versus time of activation.

by a sharp decrease of the concentration of centers in the mixture. Consequently, $O^{2-}$ basic centers observed take part in the formation of new chemical bonds Ca-O-Si and are consumed as the reaction proceeds.

The evidence for the formation calcium silicate by milling of $Ca(OH)_2$+$SiO_2$ mixture was received by cross polarized-magic angle spinning NMR [10]. $^{29}$Si NMR spectra are shown on Fig. 8.3. The peak corresponding to $Q_3$ state, representing surface silanol, predominates in starting mixture before milling, while a mixture after milling detects a predominance of the peak $Q_0$ state, i.e., with no neighbouring silicon. Whether the $Q_0$ state corresponds to tricalcium silicate ($C_3S$) or dicalcium silicate ($C_2S$) has not yet been unambiguously determined. Similar phenomena were observed on milling of kaolin [11].

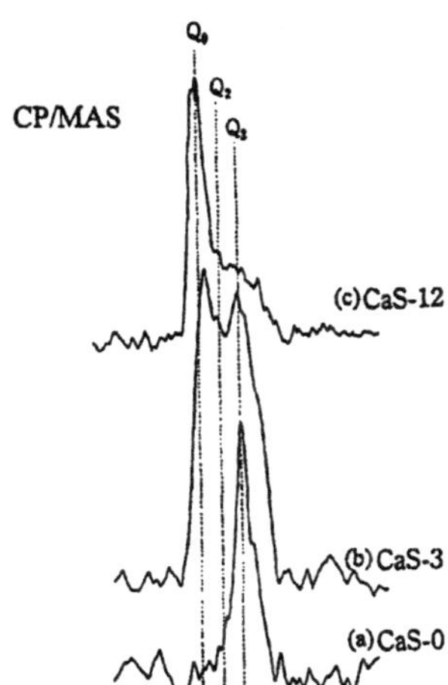


*Figure 8.3.* $^{29}$Si CP/MAS NMR spectra of physically mixed (CaS-O) and milled samples (CaS-3,12), *t* - milling time in hours.

It is known that mechanical activation generates various long-life radicals. EPR investigation were carried out for mechanically activated $Ca(OH)_2$+$SiO_2$ mixtures [12]. The mixtures were milled under vacuum. Spectroscopy measurements were carried out immediately after milling without expose to air, in order to examine the effects of ambient gaseous species. Careful analyses of the g-values and power saturation behavior, the comparison with the results of X-ray irradiation, and changes in signals after exposure to air, allowed to conclude that the main signals were attributed to $O_3^-$ radicals. On separate milling of $Ca(OH)_2$, the concentration of $O_3^-$ radicals increases.

On the other hand, the main radical species of separately milled $SiO_2$ are peroxide radicals. These two radical species, i.e., $O_3^-$ and peroxide radical, disappear on milling of a mixture of $Ca(OH)_2$ and $SiO_2$. This suggests that these two radical species are consumed by recombination to form Ca-O-Si bridging bonds by the following reaction scheme:

(i) $O_3^-$ radical formation by adsorption of an oxygen molecule onto the mechanochemically formed $O^-$ radical or $Ca\text{-}O^{\cdot}$ on the surface of partially decomposed $Ca(OH)_2$.

(ii) Formation of peroxide radicals, Si-O-O(H), by adsorption of an oxygen molecule on the radical species ≡Si.

(iii) Recombination of these two radical species to give a bridging bond such as $Ca\text{-}O^{\cdot} + \equiv Si \rightarrow Ca\text{-}O\text{-}Si$.

The relative contribution of the acid-base mechanism stated above and this radical mechanism cannot be determined individually. The concentration of the basic sites remaining after milling for 1 and 3 h is $10^7$ and $10^5$ times higher, respectively, than radical concentration. This does not straightforwardly mean, however, that the contribution of the radical mechanism is orders of magnitudes smaller, because we can measure only the concentration of remaining radical species on a separately milled sample. The amount of the radical species actually consumed by recombination has not yet been successfully determined.

The proton transfer is obtained in mechanically activated $Ca(OH)_2+TiO_2\cdot aq$ mixtures by NMR method [13]. According to NMR data, $TiO_2\cdot aq$ contains three types of oxygen-hydrogen groups [14]: (i) comparatively isolated $OH^-$ groups, (ii) $H_2O$ molecules, and (iii) three-spin systems similar to $H_3O+$ (a proton localized on water molecule). During the

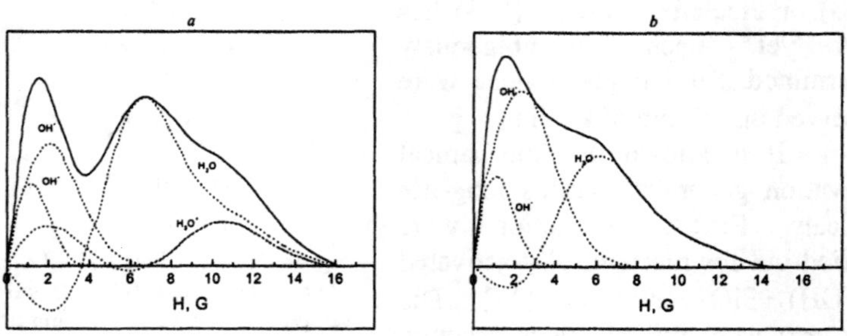


*Figure 8.4.* NMR spectra of activated $TiO_2\cdot 1.2\ H_2O$ (*a*) and $Ca(OH)_2+TiO_2\cdot 1.2\ H_2O$ (*b*).

mechanical activation of the mixture the fraction of isolated $OH^-$ group increases while the fraction of protons, constituents of the three-spin systems, sharply decreases (Fig. 8.4). Being in direct contact with $Ca(OH)_2$, protons interact with the strong basic centres to form water molecules. The resulting uncompensated bonds TiO- and -Ca produce a new bond -Ti-O-Ca-. When the concentration of water in $TiO_2{\cdot}aq$ and, consequently, the proton concentration decrease, the reaction rate decreases too.

The reaction product is X-ray amorphous; it contains water molecules and $OH^-$ groups. It crystallizes at 200÷500°C with a strong exothermic effect; the process is completed with whole dehydration (Fig. 8.5).

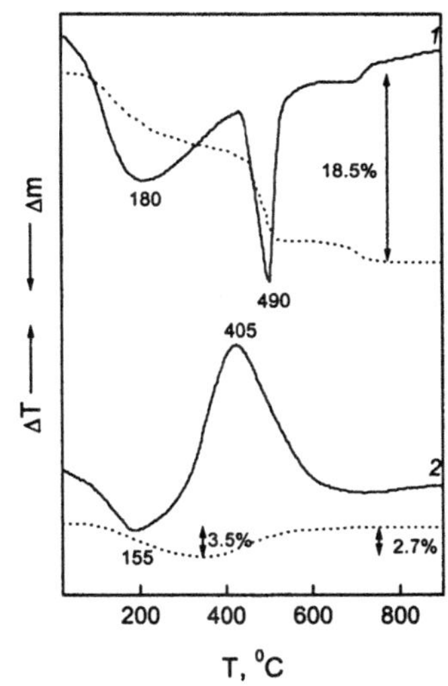


*Figure 8.5.* DTA and TG curves for the mixture $Ca(OH)_2$+$TiO_2{\cdot}1.2\ H_2O$: 1 – initial mixture; 2 – activated mixture.

According to Brönsted, the proton affinity is a quantitative characteristic for acidic and basic properties. Appropriate methods were developed for its measurement, among which is the gas adsorption on the oxide surface and the analysis of changes in vibration frequencies of the surface $OH^-$ groups [14]. The latter method was used for determination of the proton affinity for a great number of oxides, including those considered here.

Fig. 8.6 shows the difference in proton affinities ($\Delta P$=$P_{CaO}$-$P_{MxOy}$) for reactions of calcium oxide with other oxides versus the conversion rate, which was calculated from the conversion for a given time [15]. One can see that if the differences in proton affinity increase, the reaction rates increase also, as was expected. The correlation between thermodynamic Gibbs potentials of these reactions with the conversion rate is represented on Fig. 8.6 also. The data represented in Tables 2.2, 2.4, 2.5, which are ranged as acidic properties increases, can be used for determining of possibility of

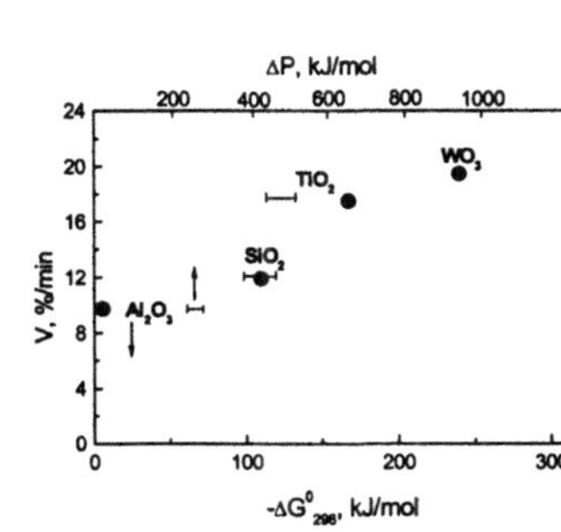


*Figure 8.6.* Correlation between the rates of mechanochemical interactions of calcium oxide with hydrated oxides and Gibbs energies for the reactions and proton affinities of anhydrous oxides.

realization of soft mechanochemical reactions in the mixtures of hydrated oxides.

The principle of electronegativity is very useful for characterization of acid-base properties and complex formation in mechanochemical reactions [16].

The most of chemical reactions accompanied by electron transfer from an atom of one reagent (reducer) to an atom of another reagent (oxidizer). Each element can have some oxidation states. The standard oxidation-reduction potential between two oxidation states of element is bonded with standard thermodynamic free energy of the transition from one state to another by the following equation:

$$\Delta G^0 = -nFE^0 \qquad (8.1)$$

where $E^0$ is a standard electrode potential of an element. It depends on the distribution of electrons in an atom. As it was pointed above, electronegativity is also the function of charge distribution, i.e. a function of the relative electron density. For this reason, a clear interdependence is observed between $E^0$ and electronegativity, as determined by the equation:

$$E^0 = 1 - 2X \qquad (8.2)$$

for the elements of main subgroups of the Mendeleev Periodic Table, and

$$E^0 = 22X - 30 \qquad (8.3)$$

for the elements of the side subgroups [17]. Here X is electronegativity (in accordance with Sanderson).

It is important to know redox potentials of elements if one wants to predict the directions of oxidative-reductive reactions between different oxidation states of elements. If $E^0$ for individual elements or semireactions are known, then the total redox potential of the system can be obtained by combining the equations that incorporate these elements or semireactions. The reactions occur when this potential is positive.

Examples of oxidative-reductive reactions under mechanical activation of systems containing hydrated oxides and any reducing agents are reported in a series of works [18-20]. For the activation of $Al+Co(OH)_2$ mixture, divalent Co ions are partially reduced to metal state and metallic aluminium is oxidized. This is confirmed by the data obtained by means of $^{50}Co$ NMR reported in [20]. The reduction can be represented by the equation:

$$2Al + 3Co(OH)_2 = 3Co + Al_2O_3 + 3H_2O$$
$$2Al + 3Co^{2+} = 2Al^{3+} + 3Co, \qquad (E = 4.07).$$

The change of the oxidation degree of atoms and ions changes their acid-base properties. For manganese compounds as example, acidity increases by factor of about 3 with oxidative degree changing by unity (Table 2.5). Therefore redox reactions can be considered also in the frames of acid-base notions described above.

## 8.2. Computational methods - DV-X$\alpha$ calculation

A simple molecular orbital calculation was made as an attempt to elucidate whether and to what extent a low coordination state plays a role on hetero-metalloxane bonds(HMB) formation [21]. First attempt aimed on $SiO_2$+$Ca(OH)_2$ system with a reasonable assumption of having silanolic OH groups on the surface of silica. Calculation was started from a model combination of $Ca(OH)_6^{4-}$ and $(SiO_4H_2)^{2-}$ clusters, shown in Fig. 8.7. They were brought into contact and the overlap population (OP) between the oxygen atom from $Ca(OH)_2$ and the hydrogen atom from silanol group was calculated. Details of the calculation methods are given elsewhere [21]. As the distance between the $Ca(OH)_6^{4-}$ and $(SiO_4H_2)^{2-}$, Z2 and Z1, respectively, decreases (Fig. 8.7), OP between the oxygen atom from $Ca(OH)_2$ and the hydrogen atom from the surface silanol, increases (Fig. 8.8).

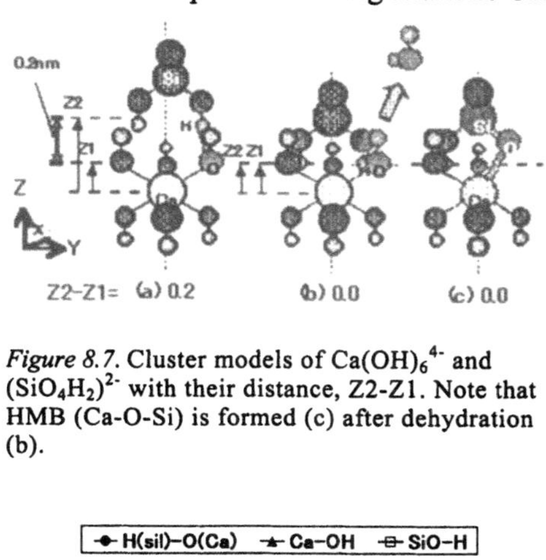


*Figure 8.7.* Cluster models of $Ca(OH)_6^{4-}$ and $(SiO_4H_2)^{2-}$ with their distance, Z2-Z1. Note that HMB (Ca-O-Si) is formed (c) after dehydration (b).

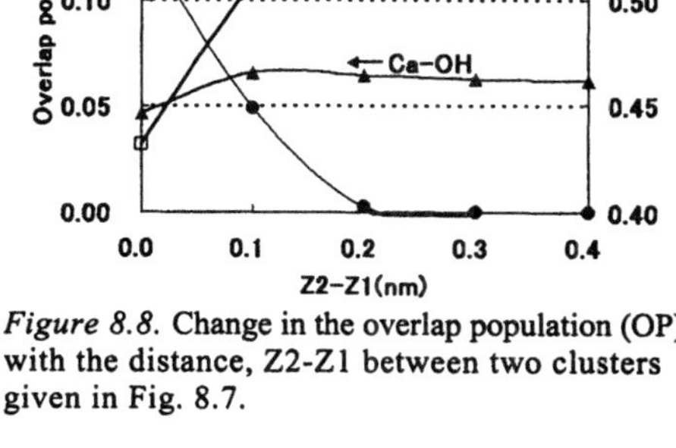


*Figure 8.8.* Change in the overlap population (OP) with the distance, Z2-Z1 between two clusters given in Fig. 8.7.

High ionicity of a Ca-O bond in a calcium oxide or hydroxide is mainly attributed to the large energy difference between the valence electrons, i.e., Ca 4s and O

2p. Madelung potential makes $O^{2-}$ levels more stable and $Ca^{2+}$ levels less stable. A Madelung constant for the particular atom generally decreases when a part of surrounding atoms disappears, i.e. with an asymmetry. With lower symmetry, hybridization becomes more likely, and dissolution of degeneration tends to occur. The energy difference between HOMO and LUMO corresponds to the band gap between oxygen 2p (HOMO), and unoccupied Ca orbitals, mainly 4s and 4p (LUMO). Thus, the effective valence levels of $Ca^{2+}$ and $O^{2-}$ come closer to each other with decreasing coordination number for Ca, as shown in Fig. 8.9. Consequently, hybridization of these Ca and O orbitals takes place in the surface state, where 4s orbitals predominantly contribute. When the surface Ca atom with lower coordination number (CN) combines with an O atom from $SiO_2$ surface, the energy levels of Ca are lowered. A smaller energy difference with respect to O 2p triggers further hybridization between the orbitals of Ca and those of O. This results in the formation of a chemical bond with higher covalency.

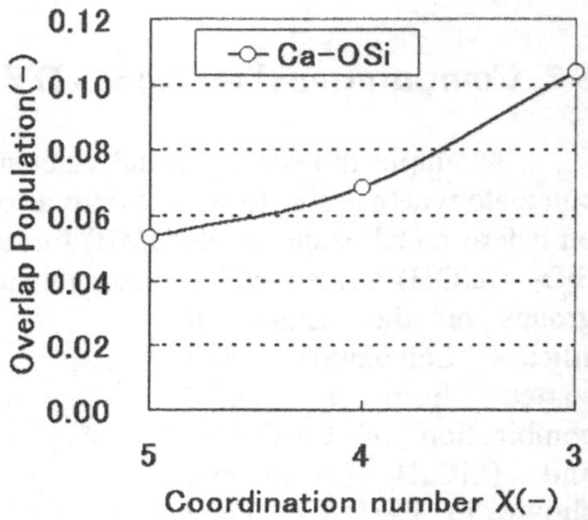


*Figure 8.9.* Change in the overlap population (OP) with the coordination number for Ca in $Ca(OH)_2$.

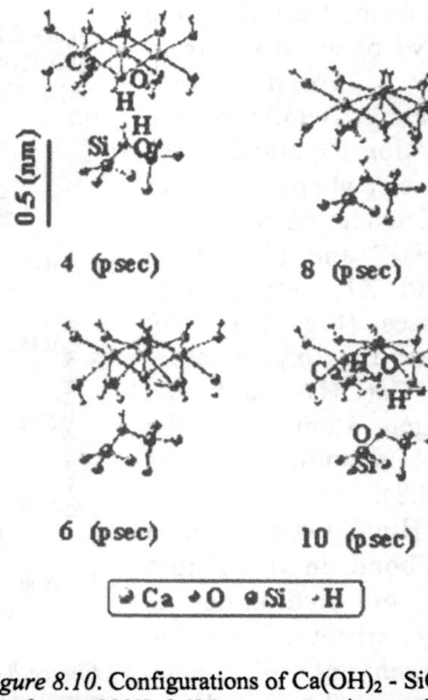


*Figure 8.10.* Configurations of $Ca(OH)_2$ - $SiO_2$ interface at 300K, 0.1Mpa, representing a scheme of proton transfer from silanolic OH to O atom of $Ca(OH)_2$.

The above mentioned results seem to be self-consistent and reasonable. However, the surface relaxation is yet to be taken into account. When a solid is excited without any chemical species to react within the close vicinity, it relaxes to restore the original, less active state, usually by a simple electron redistribution and atomic rearrangement whenever their mobility allows. Once some chemical species which are able to impinge the excited solid (in most cases, its surface) exist, then the route of relaxation may change significantly. Reactions 'in situ' may be well induced. Many examples of tribochemical 'in situ' reactions have been compiled by Heinicke [22].

Molecular dynamics (MD) simulation was used to elucidate the atomic rearrangement at the early stage of solid state mechanochemical reaction at $Ca(OH)_2$ -$SiO_2$ interface in order to take the relaxation process into account. Verlet algorithm was employed in order to calculate atomic motions [23]. Details of calculation are given elsewhere [23]. A Ca-O bond, being fairly ionic, is represented by Born-Mayer-Huggins (BMH) potential function, while for Si-O bond, an additional attractive term of the Morse potential was introduced to the BMH. Finally for O-H bond, an extra three-body-type potential was added. Inter-atomic forces from these potential energy functions within ~1.5÷2 nm around an atom was derived by using a cell model of such dimension [23].

Fig. 8.10 represents time-resolved atomic configuration at the $Ca(OH)_2$-$SiO_2$ interface versus time. Between 8 and 10 ps, a hydrogen atom from SiO-H jumps on the surface of $Ca(OH)_2$ to attach to an oxygen atom of OH group.

Surface structure of $Ca(OH)_2$ with one OH group removed is represented in Fig. 8.11. A cluster model shown in Fig. 8.11,A has no OH defect while a model with one OH defect is shown on Fig. 8.11,B. Both models, 8.11,A and 8.11,B, do not account for any kinds of surface relaxation. In contrast, a cluster model shown in Fig. 8.11,C has one OH defect with a surface relaxation. A closer look at Fig. 8.11,C reveals that low coordinated Ca atoms are displaced toward the remaining OH

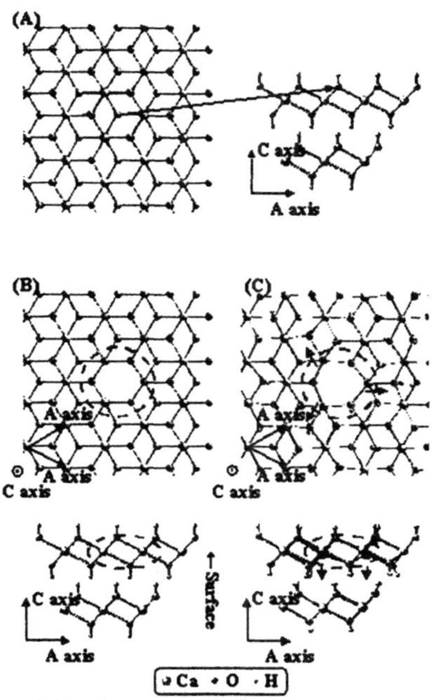


*Figure 8.11.* Cluster models for surface relaxation of $Ca(OH)_2$ due to removal of one OH group; (A) No vacancy (without surface relaxation), (B) One OH vacancy (without surface relaxation), (C) One OH vacancy (with surface relaxation).

groups, i.e., away from the OH vacancy just generated. The synchronized displacement demonstrates a surface relaxation in an usual manner toward stabilization of the electronic state of the surface. This will be examined below.

Population of Ca unoccupied orbital of non-relaxed models is shown in Fig. 8.12,A and 8.12,B, corresponding to the models shown in Fig. 8.11,A and 8.12,B, respectively. It turns out, by comparison, that the lowest energy of the unoccupied Ca orbital, pointed by arrows, lowers against the HOMO of the cluster by removing an OH group. This is in an accordance with the previous result on CaO surface, which disregards surface relaxation [21]. However, for the model accounting for the surface relaxation (Fig. 8.11,C), the energy of the unoccupied Ca orbital hardly lowers, in spite of forming of low coordinated Ca (Fig. 8.12,C).

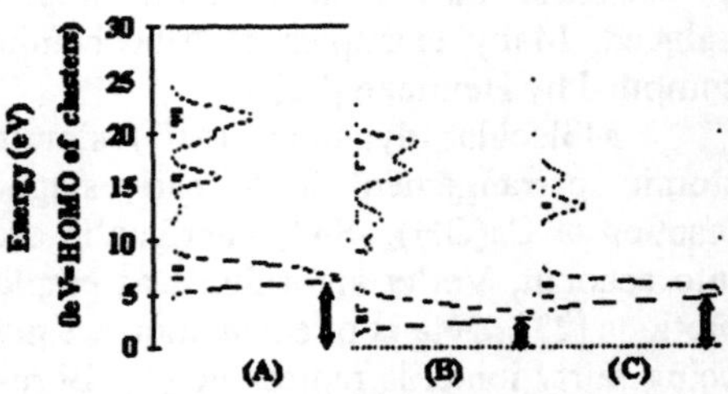


*Figure 8.12.* Orbital population of calcium unoccupied orbitals in cluster models in Fig. 4(A) ñ 4(C). (A) No vacancy (without surface relaxation), (B) One OH vacancy (without surface relaxation), (C) One OH vacancy (with surface relaxation).

Under an appreciable compressive stress, crystals deform themselves significantly. A large deviation of the resulting atomic positions from that in the ideal crystal lowers the symmetry of the crystalline lattices, and accordingly, makes a Madelung potential more asymmetric. This, in turn, forces a decrease of all the energy levels of Ca atoms to minimize the total potential energy of the reaction system. The energy difference between $Ca_{out}$ (or $Ca_{4s}$) and $O_{2p}$ decreases, while the difference between $Ca_{in}$ (or $Ca_{3p}$) and $O_{2p}$ increases. These changes intensify the bonding overlapping between $Ca_{out}$ and $O_{2p}$ toward apparent bonding molecular orbitals, while decrease the antibonding interaction between $Ca_{in}$ and $O_{2p}$. The increasing covalent nature, thus established, stabilizes a Ca-O wing of a Ca-O-Si bridging bond with respect to the other wing, O-Si, and finally, stabilizes the bridging bond as a whole.

## 8.3. Role of water in soft mechanochemical reactions

Neutralization in aqueous solutions is characterized by low activation energy and occurs very rapidly. On contrary, in the case of acid-base interaction between solids, it is multistage process and can be accompanied by the formation of stable compounds with the participation of hydrogen bonds, which are the precursors for preparation of final products by the following thermal treatment:

$$\underset{A}{M_1(OH)_n + H_mM_2O_p} \rightarrow \underset{B}{M_1(OH)_{n-2}..H..\ M_2O_p} \rightarrow \underset{C}{M_1M_2O_z + H_2O}\ .$$

When the difference between acid and base properties is not very large, the reaction can stop at the stage B with the formation of an X-ray amorphous hydrated complex compound. The formation of complex hydrated oxides is also possible, when the difference is large. There are numerous examples of the appearing of complex hydrated oxides as a result of mechanical activation of the mixtures of hydrated oxides [24-31]. $3CaO\cdot Al_2O_3\cdot 6H_2O$ and $CaO\cdot 2Al_2O_3\cdot 7H_2O$ compounds are formed in $Ca(OH)_2 + Al(OH)_3$ mixture [24,25]. A wide spectrum of hydrated calcium silicates is formed under activation of $Ca(OH)_2+SiO_2\cdot aq$ mixture [26-28]. The activation of $Ca(OH)_2+TiO_2\cdot aq$ mixture yields the compound $CaO\cdot 4TiO_2\cdot 6H_2O$ [29] while in activated $La(OH)_3+TiO_2\cdot aq$ mixture the formation of $La_2[TiO_2(OH)_2]\cdot nH_2O$ is observed [30]. The mentioned compounds are decomposed at 600-800°C with the formation of anhydrous complex oxides.

The possibility of mechanochemical reactions to occur via hydrothermal mechanism was demonstrated in [32] for the synthesis of calcium hydrosilicates. Optimal water content values (humidity) of solid mixtures of correspondent hydrates were estimated for the achievement of local hydrothermal conditions (elevated temperature and pressure) in mechanochemical activators. It was shown experimentally that the main factor determining the kinetics and composition of the products was the molar fraction of the components, the same as for hydrothermal reactions, while the effect of temperature and pressure was much weaker.

While studying the mechanism of mechanochemical synthesis of magnesium silicate, it was shown that the components are dissolved in the liquid phase and interact with each other [33]. At first, the precipitation of hydrosilicates from the solution in fine disperse and X-ray amorphous state occurs. Then the growth of larger crystals takes place. Therefore, the products are not detected directly after mechanical activation; some exposure is necessary for them to be crystallized.

It is known that the contact of two solids results in the appearance of adhesion strength due to the interaction between surface atoms. It has been stated that the interaction forces are changed with the distance between the surfaces of the condensed solids as $1/r^3$ [34,35]. Specific attraction forces are equal to 0.15 MPa at a distance of about 10 nm, 30 MPa at 3 nm, and reach 110 MPa at 1 nm. This means that at a distance of the order of atom size, van der Waals forces become very large, that helps redistribution of electrons to form a new chemical bond.

The strength of secondary formations (agglomerates), in which mechanochemical reaction occurs under mechanical action, are substantially affected by the medium used at activation. In high vacuum, the particle surface gets free from the adsorbed molecules, and the efficiency of the interaction between surface atoms increases. In the presence of liquids, especially polar ones, such as water and alcohol, molecular-dense aggregates decompose to give primary particles. That is why mechanical activation in the presence of these liquids yields finely dispersed products. The efficiency of chemical interactions between components in these systems is insignificant due to the absence of good contacts.

However, if the dispersion of a mixture of components is high and water concentration is small, so that it forms 1-3 molecular layers on the surface of solid particles, quite different phenomenon is observed. Water acts as a binding agent and improves the strength of the agglomerates [36]. The nature of intergrain contacts in these systems is determined by hydrogen and donor-acceptor bonds of water molecules with surface atoms or surface groups (e.g., OH groups) of the disperse phase.

As a rule, the surface of the disperse phase is hydroxylated (=Si-OH, =Al-OH, etc.); surface atoms possess valence d-orbitals. Therefore, the formation of adsorption water layers occurs not only via the formation of hydrogen bonds but also due to donor-acceptor bonds. It is this phenomenon that forms the grounds of wetting of solids with water.

Due to the structuring of water layer and the action of surface fields on water layers, hydrogen bonds become stronger; their energy becomes higher by a factor of 2-2.5 than that of hydrogen bonds in the water bulk, water itself exhibiting others properties ($\rho$=1.2 g/cm$^3$, dielectric permeability $\varepsilon \leq 5$) [35].

In the presence of electrolyte solutions, hydrated cations and anions participate in the formation of contacts and determine their strength. The addition of electrolytes can either increase the structuring of water layers or, vice versa, weaken this process (positive and negative hydration). Some electrolytes strengthen water structure, the others weaken it. In the first case, the viscosity of water increases; in the second case it decreases. The first group consists of polyvalent electrolytes and the electrolytes, such as halides of alkaline elements, while the second group - of some electrolytes, such as KI.

An increase of the contact strength with the decrease of water layering is confirmed experimentally. For example, it was stated that the strength of raw material formed from wetted bentonite is determined by the equation:

$$\sigma_{compress}=an^{-b}, \tag{8.4}$$

where a, b are constants, n is a number of water monolayers calculated on the basis of the given specific surface and water amount [35].

When studying the effect of the amount of water on the efficiency of interaction in the mixture of anhydrous oxides (CaO and $TiO_2$) and hydrated ones ($Ca(OH)_2$ and $H_2TiO_3$), an extremal dependence of the degree of components binding into product versus water content was found (Fig. 8.13) [37,38]. The maximum observed in this dependence can be explained from the viewpoint of the mentioned notions. At low water content, it makes the agglomerates of anhydrous oxides stronger, that helps chemical interaction to occur. At increased amount of water, the agglomerates are destroyed and chemical interaction decreases.

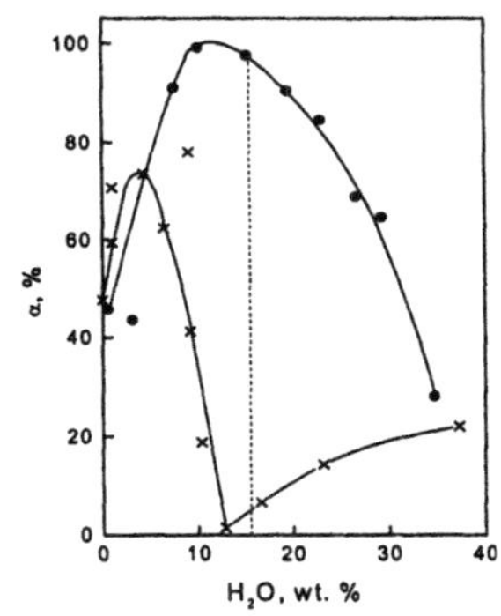


*Figure 8.13.* The degree of interaction of calcium oxide with titanium oxides as a function of water content:
1 – $CaO+TiO_2$ (anatase);
2 – $Ca(OH)_2+H_2TiO3$.

The presence of a maximum in this dependence allows to identify a new type of distinctive interactions, differing from those in the mixtures of solids and in aqueous solutions, namely, in the type of solid state mechanochemical reactions occurring at low concentration of water.

## 8.4. Mechanochemical ligand exchange as a new concept of solid state synthesis

When mechanical stress is exerted on molecular crystalline solids of coordination compounds, not only crystalline lattice but also the constituent molecules are subjected to distortion. The type and extent of distortion of each molecule is inevitably different due to the anisotropic nature of mechanical stressing. This, in turn, causes disproportionation of their ligand fields to decrease their symmetry.

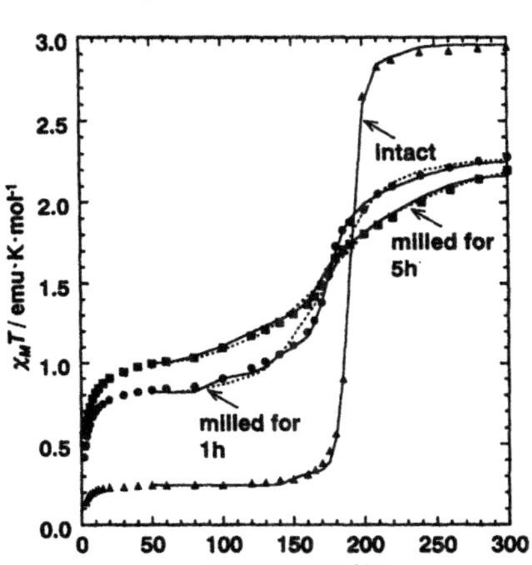


*Figure 8.14.* Tempearture dependence of observed values of $\chi_M T$ for $Fe^{II}[HB(pz)_3]_2$. Solid curves denote calculated values.

Strained molecular crystals show various anomalies in physico-chemical and electronic properties. One of the general reasons is the change of electronic states and related transition probability due to loss of the ligand field symmetry, which influences significantly on the magnetic properties and reactivity of coordination compounds [39, 40]. This is particularly important for transition metals in view of magnetic or optical properties. Partial allowance of forbidden d-d transition of $Mn^{2+}$ incorporated in ZnS, for instance, increases photoluminescence intensity [41].

When fine crystalline powders of Fe(II) coordinate compounds with octahedral symmetry are subjected to mechanical stressing, an energy level of octahedrally symmetrical 3d electrons in Fe(II) is split into two states, $e_g$ and $t_{2g}$, with a finite separation and a consequent disproportionation of the ligand field strength, $\Delta_O$ [42]. This brings about development of the distribution of $\Delta_O$. As a consequence, $3d^6$ configuration of the complex takes two different spin states, i.e., a low spin (LS) state ($^1A_{1g}$) described as $(t_{2g})^6(e_g)^0$, and a high spin (HS) one, ($^5T_{2g}$), $(t_{2g})^4(e_g)^2$.

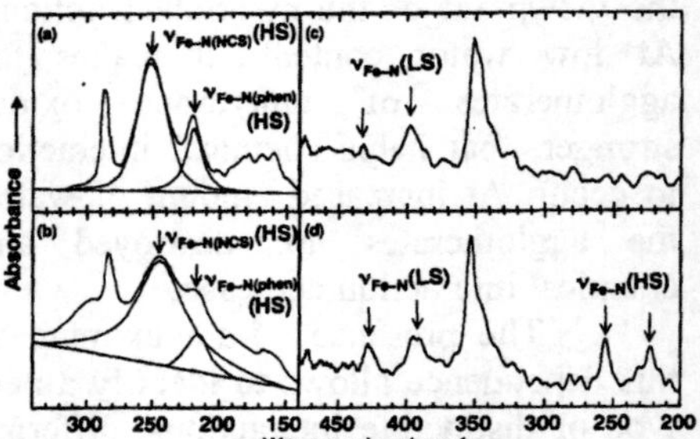


*Figure 8.15.* Far-IR spectra of $Fe^{II}(phen)_2)NCS)_2$: (a) intact, (b) after milling for 1 h and $Fe^{II}[HB(pz)_3]_2$: (c) intact, (d) after milling for 1 h.

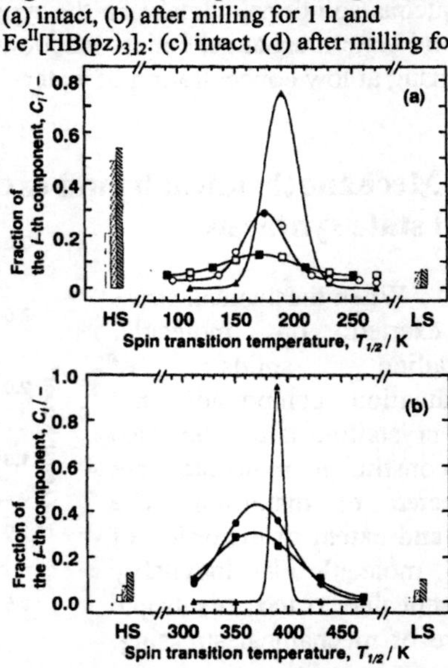


*Figure 8.16.* Variation of fraction in the *i*-th component with spin transition temperature for (a) $Fe^{II}(phen)_2$; (b) $Fe^{II}[HB)pz)_3]_2$. Intact; after milling for 1 h; after milling for 5 h.

The change in the spin crossover into multi-mode was actually observed by mechanical stressing exerted to their crystalline particles of the complex compounds, $Fe^{II}(phen)_2(NCS)_2$ (phen = 1,10-phenanthroline) and $Fe^{II}[HB(pz)_3]_2$ (pz = pyrazol-1-yl) [43].

The spin crossover behavior of $Fe^{II}(phen)_2(NCS)_2$ is displayed in Fig. 8.14 where $\chi_M T$ is plotted against temperature. Along with the increase at around 2 ~ 50 K as was already found for several other compounds [39], the change in $\chi_M T$ with temperature around 190 K becomes flatter after milling.

Curves (a) and (b) in Fig. 8.15 show far-IR spectra of $Fe^{II}(phen)_2(NCS)_2$. Fe-N(NCS) and Fe-N(phen) stretching peaks appear at 249 $cm^{-1}$ and 218 $cm^{-1}$, respectively [44-47]. Increase in the peak width, $\Delta W$, relative to the intact sample is particularly remarkable for Fe-N(NCS).

Fractional distributions of the spin crossover component, $c_i$, for $Fe^{II}(phen)_2(NCS)_2$ and $Fe^{II}[HB(pz)_3]_2$ are shown in Fig. 8.16*a*. Since spin transfer probability varies with $|\Delta - P| \sim k_B T$, the vertical temperature axis corresponds linearly to the ligand field strength, $\Delta$. Fig. 8.16*b* displays that the consequence of milling is the decrease of $\Delta$ in the main spin crossover component and the broadening of the distribution of the ligand field strength. Note, that the broadening takes place in both directions, i.e., with increasing and decreasing energy. Furthermore, the broadening of $\Delta$ in both directions also brings about residual HS states, $c_{HS}$, and residual LS states, $c_{LS}$. These results demonstrate the disproportionation of the ligand field strength, corresponding to the multi-mode spin crossover.

The increase in the FWHM of the Mössbauer spectra was observed for both complexes. This is another indication of the ligand field disproportionation by milling, even without changing the spin states from LS to HS.

Increases in $\Delta W$ shown in Figs. 8.17*a* and 8.17*b* indicate the broadening of the distribution in the Fe-N bond energy by milling of $Fe^{II}(phen)_2(NCS)_2$. This is remarkable particularly for Fe-N(NCS) stretching peak of the HS state. This further indicates that Fe-N coordination bond length scatters wider after milling.

Potential energy diagram for LS and HS states of $Fe^{II}(phen)_2(NCS)_2$ at room temperature is illustrated schematically in Fig. 8.17*a*. Energy levels in the potential energy curve

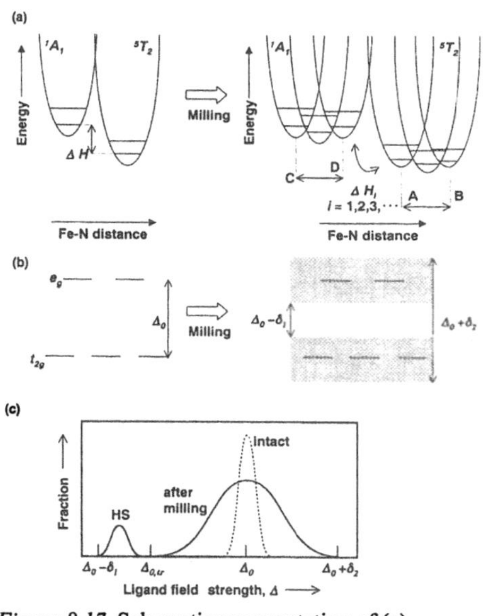


*Figure 8.17.* Schematic representation of (a) potential energy diagram for LS and HS states of $Fe^{II}(phen)_2(NCS)_2$ at room temperature, (b) energy gap distribution between $t_{2g}$ and $e_g$, $\Delta$, of $Fe^{II}(3d^6)$, (c) the distributio of ligand field strengh.

stand for the ground and the exited states of vibrational energy. The broadening of the Fe-N bond length distribution, A ~ B, corresponds to the plurality of $^5T_2$ potential energy curves. Only three different potential energy curves were illustrated in Fig. 8.17*a*. The number of curves (three - in this particular case) does not play a significant role. Similar broadening, C ~ D, and corresponding the plurality of $^1A_1$ potential energy curves may be represented in the complex molecules of the LS state as well.

Loss of ligand field symmetry leads further to a number of solid state reactions unique for coordination compounds. As a mixture of $FeCl_2{\cdot}4H_2O$ with 1,10-phenanthroline (phen) was milled, a complex $[Fe(phen)_3]Cl_2$ with an apparent yield close to 100% was obtained [47]. This is an example of ligand exchange reaction. Unlike the case of liquid phase reaction, an exchange reaction is triggered by elimination of one water molecule coordinated to Fe(II). Based on a series of analyses, including X-ray photoelectron spectroscopy, it was concluded that the solid state ligand exchange is attributed to the deformation of the ligand field and consequent disproportionation of the coordination bonding states of 4 water molecules, as schematically shown in Fig. 8.18. Once a water molecule is deleted in the presence of a species which stabilizes the ligand field, such as phen, its one leg makes easily a coordination bond to Fe(II). This facilitates further exchange with the second water molecule to complete the coordination of the first phen molecule. Similar replacement proceeds until all the four water molecules are replaced by two phen molecules, since the asymmetry of the ligand field increases by a partial ligand exchange reaction and, hence, enhances further exchange.

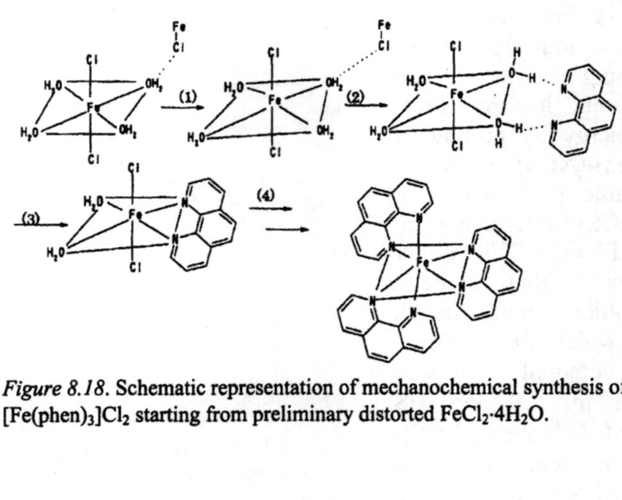


*Figure 8.18.* Schematic representation of mechanochemical synthesis of $[Fe(phen)_3]Cl_2$ starting from preliminary distorted $FeCl_2{\cdot}4H_2O$.

### *References*

1. Watanabe T., Liao J.F., Senna M. Changes in the surface properties of $Me(OH)_2$ and $Me(OH)_2$-$SiO_2$ (Me: Ca, Mg, Sr) due to mechanical activation. J. Solid State Chem. 1995; 115: 390-94.
2. Watanabe T., Isobe T., Senna M. Mechanisms of incipient chemical reaction between $Ca(OH)_2$ and $SiO_2$ under moderate mechanical stressing. I. A solid state acid-base reaction and charge transfer due to complex formation. J. Solid State Chem. 1996; 122: 74-80.
3. Liao J.F., Senna M. Mechanochemical dehydration and amorphization of hydroxides of Ca, Mg and Al on grinding with and without $SiO_2$. Solid State Ionics 1993; 66: 313-19.
4. Watanabe T., Liao J., Senna M. Changes in the basicity and species on surface of $Me(OH)_2$-$SiO_2$ mixtures due to mechanical activation. J. Solid State Chem. 1995; 115: 390-95.
5. Paukshtis, Evgenii, *Infrared spectroscopy in heterogeneous base-acid catalysis*. Novosibirsk, Nauka, 1992.
6. Kosova N.V., Devyatkina E.T., Avvakumov E.G. Surface basic and acidic species and mechanochemical reactions in mixtures of hydrated oxides. Doklady Akad. Nauk 1996; 347: 388-92.
7. Kosova N.V., Devyatkina E.T., Avvakumov E.G. et al. Mechanochemical reactions of hydrated oxides: some aspects of mechanism. Chem. Sustainable Development, 1998; 6: 125-30.
8. Kosova N.V., Asanov I.P., Devyatkina E.T., Avvakumov E.G. Investigation of atom state change in surface layer at the initial stages of mechanochemical reactions by means of X-ray electron spectroscopy. Zhurn. Neorg. Khimii 1999; 44: 100-2.
9. Hess W. Dissertation, Tech. University of Karlsruhe, Karlsruhe 1980.
10. Watanabe T., Isobe T., Senna M. Mechanisms of incipient chemical reaction between $Ca(OH)_2$ and $SiO_2$ under moderate mechanical stressing. III. Changes in the short range ordering through the mechanical and thermal processes: J. Solid State Chem. 1997; 130: 284-89
11. Nakano T., Kamitani M., Senna M. Mechanical activation of kaolin. Mater. Sci. Forum 1996; 225-227: 587-91.
12. Watanabe T., Isobe T., Senna M. Mechanisms of incipient chemical reaction between $Ca(OH)_2$ and $SiO_2$ under moderate mechanical stressing. II. Examination of a radical mechanism by an EPR study. J. Solid State Chem. 1996; 122: 291-96.
13. Pletnev R.A., Ivakin A.A., Kleschev D.G., *Hydrated oxides of elements of IV and V groups*. Moskva: Nauka, 1986.
14. Kosova N.V., Devyatkina E.T., Denisova T.A., Avvakumov E.G. Proton transfer in mechanochemical reactions of hydrated oxides. Zhurn. Neorg. Khimii 1999; 44: 912-16.
15. Avvakumov E.G., Devyatkina E.T., Kosova N.V. Mechanochemical reactions of hydrated oxides. J. Solid State Chem. 1994; 113: 379-83.
16. Watanabe T., Isobe T., Senna M. Electronegativity equalization during mechanochemical addition reaction. Chem. Sustainable Development 1998; 6: 165-170.
17. Eggins B.R., *Chemical Structure and Reactivity*. London: Macmillan, 1976.
18. Kojima Y., Isobe T., Senna M., Shinohara T., Ono S., Sumiyama K., Suzuki K. Mechanism of complex formation between metallic Al and titania hydrogel via a mechanical route. J. Mater. Res. 1996; 11: 1305-9.
19. Kojima Y., Isobe T., Senna M., El-Eskandarany M.S., Sumiyama K., Suzuki K. Comparison of mechanical complexation and thermal behavior of metal, oxide and hydrogel mixtures of the Al-Ti-O system. Int. J. Mechanochem. Mech. Alloying 1994; 1: 32-41.

20. Stone F.S., Garrone E., Zecchina A. Surface properties of alkaline earth oxides as studied by UV-visible diffuse reflectance spectroscopy. Mater. Chem. Phys. 1985; 13: 331-46.
21. Fujiwara Y., Isobe T., Senna M., Tanaka J. Effects of reduced coordination number for Ca on the electron redistribution during Ca-O-Si bridge bonding from CaO or $Ca(OH)_2$ and $SiO_2$. J. Phys. Chem. 1999; 103: 9842-46.
22. Heinicke G. *Tribochemistry*. Berlin:Akademie-Verlag, 1984.
23. Verlet L. Computer "experiments" on classical fluids. I. Thermodynamic properties of Lennard-Jones molecules. Rhys. Rev. 1967; 159: 98-103.
24. Kvitkovsky A.K., Kosova N.V, Avvakumov E.G. et al. Mechanochemical synthesis of calcium hydroaluminates and their utilization for preparation of binder from slages of aluminium melting. Chem. Sustainable Development, 2000; 8: 641-45.
25. Temuujin J., Mackenzie K.I.D. Effect of mechanochemical treatment on the synthesis of calcium dialuminate. J. Materials Chemistry 2000; 10: 1019-23.
26. Kosova N.V., Devyatkina E.T., Avvakumov E.G. Mechanochemical synthesis of calcium silicates on the base of hydrated oxides. Sibirskii Khim. Zhurn. 1992; 2: 135-43.
27. Mi G., Saito F., Hanada M. Mechanochemical synthesis of tobermorite by wet grinding in a planetary ball mill. Powder Technology 1997; 93: 77-81.
28. Sasaki K., Masuda T., Ishida H., Mitsuda T. Synthesis of calcium silicate hydrate with Ca/Si=2 by mechanochemical treatment. J. Am. Ceram. Soc. 1996; 80: 472-76.
29. Kosova N.V., Avvakumov E.G., Malakhov V.V. et al. About nature of phases formed in soft mechanochemical synthesis of calcium titanate. Doklady Akad. Nauk, 1997; 356: 350-53.
30. Shlyahtina A.V., Sherbakova L.G., Kolbanev I.V. Synthesis of rare earth dititanates using fine-grained reagents. In Abstracts of 3$^{rd}$ INCOME, September 4-8, 2000, Prague, Chekhiya.
31. Liao J.F., Senna M. Crystallization of titania and magnesium titanate from mechanically activated $Mg(OH)_2$ and $TiO_2$ gel mixture. Mater. Res. Bull. 1995; 30: 385-92.
32. Boldyrev V.V., Khabibullin A.Kh., Kosova N.V., Avvakumov E.G. Hydrothermal reactions under mechanical activation. J. Synthesis and Processing, 1996; 4: 371-81.
33. Temuujin J., Okada K., Mackenzie K.J.D. Formation of layered magnesium silicate during the ageing of magnesium hydroxide-silica mixtures. J. Am. Ceram. Soc. 1996; 81: 574-76.
34. Deryagin B.V., Krotova N.A, Smilga V. P., *Adhesion of solids*. Moskva: Nauka, 1973.
35. Sichev M.M. Modern conceptions of the nature of cement components inorganic binders and the role of polymer formations. In *Chemistry and technology of silicates and refractory materials*. Institute of silicate chemistry of I.V. Grebenschikov Leningrad: Nauka, 1989.
36. Karyakin A.V., Kriventsova T.A., *Water state in organic and inorganic compounds*. Moskva: Nauka, 1971.
37. Avvakumov E.G. Soft mechanochemical synthesis as a basic for new chemical processes. Chem Sustainable Development 1994; 2: 541-59.
38. Avvakumov E.G., Kosova N.V. Soft mechanochemical synthesis: specific features and outlook. Chemistry Reviews. Harwood Academic Publishers. 1998; 23 (part 2): 285-312.
39. Tsuchiya N., Tsukamoto A., Ohshita T., Isobe T., Senna M, Yoshioka N., Inoue H. Anomalous spin crossover of mechanically strained iron ($\alpha$) complexes with 1,10-phenantroline with their counter ions, NCS- and PF6-. J. Solid State Chem. 2000; 153: 82-91.
40. Tsuchiya N., Isobe T., Senna M., Yoshioka N., Inoue H. Mechanochemical effects on the structures and chemical states of [ $Fe(Phen)_3](NCS)_2{\cdot}H_20$. Solid State Commun. 1996; 99: 525-29.

42. Tsuchiya N., Tsukamoto A., Oshita T., Isobe T., Senna M., Yoshioka N., Inoue H. Stress-induced ligand field distribution and consequent multi-mode spin crossover in $Fe^{II}(phen)_2(NCS)_2$ and $Fe^{II}[HB(pz)_3]_2$. Solid State Sciences, to be published.
43. Adams D.M., Long G.J., Williams A.D. Spectroscopy at very high pressures. An infrared study of spin-state equilibrium in some iron (II) complexes. Inorg. Chem. 1982; 21: 1049-53.
44. Takemoto J., Hutchinson B. Effect of magnetic crossover on the low-frequency infrared spectrum of diisocyanatobiz (1, 10-phenanthroline)-iron. Inorg. Nucl. Chem. Lett. 1972; 8: 769-72.
45. Takemoto J., Hutchinson B. Low frequency infrared spectra of complexes which exhibit magnetic crossover. I. Iron (II) complexes of 1,10-phenanthroline and 2,2-bipyridine. Inorg. Chem. 1973; 12: 705-8.
46. Oshita T., Nakajima D., Tsukamoto A., Tsuchiya N., Isobe T., Senna M., Yoshhioka N., Inoue H. Mechanisms of mechanochemical solid-state synthesis of tris (1,10-phenanthroline) iron (II) chloride complex. To be published.

# Chapter 9

# INDUSTRIAL APPLICATIONS OF SOFT MECHANOCHEMICAL SYNTHESIS

## 9.1. The preparation of ceramic concrete based on cement-free ceramic binding systems

Mechanical activation of oxides in the presence of water which result in the formation of hydrated forms able to be polymerized has been developed as a method for the preparation of aqueous ceramic binding suspensions (ACBS). ACBS undergo spontaneous structuring and binding, this providing the formation of final products with compression strength of 20–36 MPa. One of the first publications on this process appeared in 1936 [1]. The method was developed first by Pivinsky [2-9] and then by other researchers [10-13].

The effect of substantial binding properties exhibited by ceramic suspensions was discovered for vitreous quartz [2,3]. Ingots, prepared from high-concentrated suspensions of vitreous quartz and dried, exhibited tensile bending strength of 3÷7 MPa and compression strength 70÷90 MPa, due to low porosity and gluing ability of silicic acid. In 1979, ACBS were prepared using quartz sand which was traditionally considered as inactive substance [5]; later, they were prepared from other oxides: $Al_2O_3$, $ZrO_2$, $MgO$, $ZrSiO_4$, $3Al_2O_3{\cdot}2SiO_2$, etc. [6,7].

According to [12], one of the reasons why binding properties appear in quartz sand and aluminosilicates is the accumulation of silicic or aluminosilicic acids and their polymerization during activation followed by stabilization (intensive stirring). The concentration of silicic acid in the suspension increases during treatment in ball mill to 0.5 g/l in neutral medium and to 20.5 g/l in alkaline medium.

The strength characteristics of semi-products obtained from the suspensions are also higher for the suspensions activated in alkaline media (pH=9.4) and equal to 38-60 MPa, comparing with 9-14 MPa for suspensions activated in neutral medium [13]. An increase of the temperature of suspension during activation also increases hydration and helps strength increasing. In order to increase the strength of semi-products as prepared, they are thermally treated at 105-110°C or impregnated with alkaline solutions.

Different methods are used to prepare water-ceramic binding suspensions, depending on initial reagents. Wet grinding (followed by stabilization) is optimal for silica and aluminosilicate materials, while dry grinding and addition them into the suspension are sufficient for magnesia compositions.

Prediction of the properties of suspensions is based on the consideration of the chemical properties of oxides. As it was noticed above (Chapter 2), one of the parameters that characterize acid-base properties of an ion, is ion potential (charge to radius ratio). A correlation that was found in [8] between ion potential and maximal achieved technological concentration of solid oxide in water-ceramic suspension is shown in Table 9.1.

*Table 9.1.* Correlation between ion potential of the cation of oxide and maximal volume concentration of the solid phase in water-ceramic suspension [8].

| | Vitre-ous quartz | Quartz sand | High-alumina chamotte (64.2% mass.$Al_2O_3$) | Synth. mullite $3Al_2O_3{\cdot}2SiO_2$ | $Al_2O_3$ calcined at 1000°C | $ZrO_2$ natur. | MgO electro-fused |
|---|---|---|---|---|---|---|---|
| Ion potential (Z/r) | 10 | 10 | 6.70 | 6.46 | 5.30 | 4.88 | 2.70 |
| Max. vol. concentration of oxide | 0.80 | 0.75 | 0.72 | 0.70 | 0.62 | 0.58 | 0.50 |

The preparation of ceramic products and ceramic concrete using water-ceramic suspensions has been widely used in the production of constructive materials [9,14].

## 9.2. Improvement of cement binders

Cement binders are based on silicates, aluminates, aluminium silicates, aluminium ferrites of alkaline earth metals (mainly calcium). The hydration of these compounds or their mixtures with the formation of hydrated compounds is a process which provides binding and the preparation of constructive products.

Mechanical activation has not yet found application for the synthesis of the main minerals of cement; however, it is used to increase the activity of cement clinker [15-19].

Promising technologies for cement preparation are those using wastes. The above-mentioned compounds are present in ash formed by coal combustion, in the wastes of casting and abrasive production, etc. The quality of binders prepared from these materials depends on the correct selection of the mixture composition (it should correspond to the composition of the cement) and on proper mixing providing the components to give hydrated compounds when coming in contact with water [20]. It was shown that mechanical activation of mixtures of different industrial wastes with composition close to that of cement, allows substantially increasing their binding properties [21].

However, as we have demonstrated above, mechanochemical interaction between anhydrous oxides is insufficiently effective. But when a small amount of water is added to a mixture of anhydrous oxides, one can observe the intensification of reactions since more reactive hydrated oxides are formed.

This idea was realized in [22] where a composed binder was obtained from industrial wastes mechanically activated in the presence of small amounts of water. An initial mixture for binder preparation included high-calcium ash (the product of coal combustion at heat plant), worked-out mixture of casting industry containing silica as a main component, and high-alumina waste of the abrasive industry, i.e. all cement-forming compounds, as well as the products of their interaction. This mixture was activated in a planetary mill of periodic action in the presence of small amounts of water (~2-5 % mass). This method provides the decrease of the amount of unbound calcium oxide that causes worsening of long-time strength characteristics of product in water. The properties of the binder thus prepared were similar to those of standard. The results are shown in Table 9.2.

Calcium-aluminate cements containing $CaO{\cdot}Al_2O_3$, $12CaO{\cdot}7Al_2O_3$, $CaO{\cdot}2Al_2O_3$, etc. exhibit good binding properties. They are used to prepare fast solidified, high-strength and fire-proof concrete. Being added to ordinary ones, they improve the characteristics of the latter providing the accelerated solidification and the increased strength.

The authors [23] demonstrated that a substantial effect could be achieved by adding a mixture of separately activated $Al(OH)_3$ and $Ca(OH)_2$, taken at a molar ratio of 1:1, to ordinary cement instead of calcium aluminates. It was also found that the mechanical activation of only $Al(OH)_3$ was enough, since $Ca(OH)_2$ exhibits substantial solubility in water.

*Table 9.2.* Changes of specific surface area, free calcium oxide content, strength and water resistance of composed binder product prepared from a mixture of mechanically activated industrial wastes versus activation time [22].

| Activation time, min | Dispersion of mixture, $m^2/kg$ | CaO content, % | Compression strength, MPa | | Softening coefficient | Water-proof properties |
|---|---|---|---|---|---|---|
| | | | $R_{dry}$ | $R_{sat. sol.}$ | | |
| 0 | 200 | 12.6 | 27.82 | 6.36 | 0.49 | Low |
| 1 | 400 | 6.33 | 35.63 | 9.87 | 0.57 | // |
| 3 | 500 | 4.82 | 41.35 | 25.73 | 0.88 | // |
| 6 | 600 | 3.92 | 52.48 | 47.65 | 1.59 | High |
| 10 | 750 | 2.80 | 56.75 | 49.98 | 1.85 | // |
| 15 | 1000 | 2.02 | 43.54 | 40.54 | 1.34 | // |

The above examples show that the application of mechanical activation to the improvement of the properties of binders, as well as the use of industrial wastes in the preparation of binders, appear to be very promising.

## 9.3. Preparation of sorbents, catalysts and their supports

**Sorbents.** At present, large amounts of phosphate raw material is processed to manufacture phosphorous acid and fertilizers. Fluorine that is present in raw materials evolves into gaseous phase during treatment. It is partially captured by wetted absorbers; however, a significant part of it is released into the atmosphere. Since it is a valuable product, very strict regulations are applied to fluorine emission. The increase of its utilization degree is considered to be a very urgent problem.

Since wet purification methods are insufficiently effective, solid sorbents are used. The systems based on calcium and aluminium oxides and hydroxides are promising as such sorbents. A sorbent should provide a minimum gas dynamic resistance, therefore, it is produced as the granules of complicated shape. Sorbents based on lime and aluminium hydroxide are good fluorine absorbers. However, lime sorbents do not possess sufficient mechanical strength, while the use of aluminium hydroxide alone as a sorbent is limited by its high cost.

The authors [24,25] proposed to use lime as a raw material and aluminium hydroxide as reinforcing additive. It was proposed to treat this mixture by mechanical activation. As it was shown above (see Chapter 6), $3CaO{\cdot}Al_2O_3{\cdot}6H_2O$ is formed during the activation of such a mixture. Thus, the presence of calcium hydroxide provides good absorbing ability towards fluorine while aluminium hydroxide is responsible for mechanical strength of granules, due to formation of reinforcing calcium hydroaluminates. In order to increase the amount of $3CaO{\cdot}Al_2O_3{\cdot}6H_2O$, the granules are treated hydrothermally with sharp vapour. The absorber prepared from mechanically activated mixture of $Ca(OH)_2$ and $Al(OH)_3$ at a molar ratio of 1:1 is characterized by the porosity of 56 vol. %, mechanical strength of 11.1 MPa, and provides the absorption of fluorine and its compounds at a level of 98-99 %, phosphorous acid of 70 % during 150 h operation time under industrial conditions [25].

Because of increasing consumption of lithium compounds, the search for new technologies of lithium recovery from pickles, enriched with calcium and magnesium chlorides, is a very urgent problem. It has become possible to solve this problem with the help of a sorbent based on chlorine-containing form of a double lithium and aluminium compound, $LiCl{\cdot}2Al(OH)_3{\cdot}mH_2O$, with lithium deficit. This compound was prepared by soft mechanochemical synthesis from $Al(OH)_3$ and $LiCl{\cdot}H_2O$ [26,27]. The use of a blade mixer was sufficient to provide this reaction [27]. When fine $Al(OH)_3$ was used, the synthesis time was equal to 10 min. This allows to prepare the product at rather high productivity. Specific surface area of $LiCl{\cdot}2Al(OH)_3{\cdot}mH_2O$ as prepared was 4.9÷6.2 $m^2/g$, the static exchange capacity with respect to lithium was 8.0 mg/g, while theoretical value being 8.8 mg/g. Thus synthesized powder was used to prepare granules with well-developed porosity (~0.2÷0.3 $cm^3/g$).

A method and apparatus for selective sorption extraction of lithium from liquors (with total salt content of ~350 g/l) have been developed [29-31]. The method involves the treatment of granulated $LiCl{\cdot}2Al(OH)_3{\cdot}mH_2O$ compound with pickle till complete saturation with lithium under the conditions of counter-flow regime, followed by lithium desorption from the sorbent. The working capacity of the sorbent was 6.3 mg of lithium per 1 g of sorbent, that corresponds to 90 % of its initial capacity. The degree of lithium recovery under counter-flow regime corresponds to 94-99 %. The resulting solutions contain lithium chloride (3.1-3.5 g/l), with minor impurities of sodium, magnesium and calcium salts. Concentrated lithium chloride solutions were used to precipitate lithium carbonate with soda [32,33].

**Catalysts and supports.** The oxidation of ammonia by the oxygen of air is known to be the main industrial process for nitric acid production. The process includes three stages: the oxidation of $NH_3$ into NO by the oxygen of air in the presence of catalysts $4\ NH_3 + 5\ O_2 = 4\ NO + 6\ H_2O$, the oxidation of NO into $NO_2$ in oxygen excess 2 NO + $O_2$= 2 $NO_2$, and the interaction of nitrogen oxides with water 3 $NO_2$ + $H_2O$ = 2 $HNO_3$ + NO.

At present, the main industrial catalyst of ammonia oxidation is platinum and its alloys with aluminium and rhodium. Taking into account the deficit and high cost of platinum metals, the decreasing of the consumption and losses of platinum metals is an urgent problem. Therefore, several compositions of complex oxide catalysts have been developed with iron (III), cobalt and chromium oxides as an active component. Complex oxides with perovskite structure are used as new catalysts; they provide selective oxidation of ammonia with an yield not less than 90 %. The authors of [33] proposed to use perovskite powders $LaMeO_3$, where Me=Fe, Co, Ni, Cr, Mn, and $La_{1-x}Sr_xMeO_3$, where Me=Co, Mn and x=0.25-0.75. To prepare these compounds, they used the precipitation by tetraethyl ammonia from diluted nitrate solutions taken at necessary ratios. The powders as prepared are poorly molded as in the form of honeycomb structures as well as in the form of simple granules.

The authors [34] proposed to use perovskites $ABO_3$, where A are calcium cations, or a mixture of calcium and lanthanum, and B are iron, cobalt, nickel or manganese cations, or their mixtures. Besides, aluminates, silicates, aluminium silicates, zirconates and chromates of different types are added as structure-forming components providing strength and stability to thermal shocks [34].

The proposed block catalysts of honeycomb structure are characterized by the following component ratio, mass. %: perovskite – 50-95, the oxides of aluminium or silicon, zirconium, chromium 5-50. A distinguishing feature of the catalyst is the use of perovskites $CaMeO_3$ and $Ca_xLa_{1-x}MeO_3$, where Me=Fe, Co, Ni, Mn and x=0.9-0.1, and of active components with structure-forming and thermostabilizing properties, such as aluminium, silicon, zirconium, chromium compounds.

Complex oxides with perovskite structure are prepared by means of mechanochemical synthesis. A mixture of initial oxides, hydroxides or carbonates are subjected to mechanical activation, then the powder is calcined at 600-800°C for 2-4 h. The specific surface area of thus prepared perovskites is 10-20 $m^2/g$ [35-37].

The tests of catalyst stability towards long operation under industrial conditions showed that the catalyst remained highly active after three months and its strength being high. Besides, it was showen that the use of ammonia oxidation catalyst of a regular structure as the second step of the catalytic system allows to decrease the investment and the loss of platinum metals by 30 % and 20 %, respectively. Due to the regular structure, the proposed catalysts improve gas dynamics of the process; it becomes more stable, the breakthrough of ammonia is completely excluded, that improves the safety of the process. The catalyst is stable towards sharp changes of temperature.

Complex oxide compounds, such as mullite ($3Al_2O_3 \cdot 2SiO_2$), cordierite ($2MgO \cdot 2Al_2O_3 \cdot 5SiO_2$), celsian ($BaO \cdot Al_2O_3 \cdot 2SiO_2$), anortite ($CaO \cdot Al_2O_3 \cdot 2SiO_2$), thialite ($Al_2TiO_5$), zircon ($ZrSiO_4$), are used as catalyst supports. Their preparation has been described above (see Chapter 7).

## *References*


1. Pines B.Ya., Ter-Mikhaelyants E.I. Artificial clay-like state of high-fire-proof materials. Ogneupory 1936; 3: 74-84.
2. Pivinsky Yu. E., Gorobets F.T. High-density quarts ceramics. Ogneupory 1968; 8: 45-50.
3. Pivinsky, Yurii, Romashin, Alexandr, *Quartz Ceramics*. Moskva: Metallurgia, 1974.
4. Pivinsky Yu.E. The foundations of ceramoconcrete technology. Ogneupory 1978; 2: 34-9.
5. Pivinsky Yu.E., Bevz V.A., Mityakin P.L. The basic principles of the preparation of high-concentrated suspensions of quartz sand. Ogneupory 1979; 3: 46-51.
6. Pivinsky Yu.E., Bevz V.A. The preparation of aqueous zircon suspensions and investigation of their rheological and technical properties. Ogneupory 1979; 8: 38-43.
7. Pivinsky Yu.E., Bevz V.A. The preparation of aqueous mullite suspensions and investigation of their rheological and technical properties. Ogneupory 1980; 3: 45-50.
8. Pivinsky, Yurii, *Ceramic Binders and Ceramoconcrete*. Moskva: Metallurgia, 1990.
9. Pivinsky, Yurii, *New Fire-Proof Concrete*. Belgorod: BGTASM, 1996.
10. Mityakin, Pavel, Rozental', Oleg, *Heat-Proof Materials Based on Aqueous Ceramic Binding Suspensions*. Novosibirsk: Nauka, 1987.
11. Mityakin P.L., Rozental' O.M., Pletnev R.N. Adhesion ability of the hydrate coating in the system quartz – water. Izvestia AN SSSR: Neorg. Materialy 1983; 5: 760-63.
12. Mityakin P.L. The main principles of mechanochemical preparation of heat-proof binding systems and materials based on them. Izvestia SO RAN, ser. khim. nauk 1985; 5: 119-23.
13. Mityakin P.L., Solomin N.V. About solodification and strength characteristics of a binder made of quartz sand. Izvestia AN SSSR: Inorganic materials 1981; 17: 1102-5.
14. Pivinsky Yu.E., Trubitsyn M.A. Fire-proof concrete of the new generation. Cement-free concrete. Ogneupory 1990; 8: 6-10.
15. Juhasz, Z. and Opoczky, L., *Mechanical Activation of Silicates by Fine Grinding*. Budapest: Akad. Kiado, 1982.
16. Juhasz, A. and Opoczky, L., *Mechanical Activation of Minerals by Grinding: Pulverizing and Morphology of Particles*. Budapest: Akad. Kiado, 1990.
17. Solomatov V.I., Guseva A.Yu., Kononova O.V., Kononov M.Yu. The effect of mixed binder grinding method on the formation of strength of cement compositions. Beton i zhelezobeton 1999; 1: 5-6.
18. Seculis Z., Popov S., Milosevic S. Comminution and mechanical activation of portland cement in different mill types. Ceram. Silikaty 1998; 42: 25-28.
19. Sulimenko L.M., Krivoborodov Ju.R. Influence of mechanical activation of raw on clinkerformation processes and properties of cement. Zhurn.Prikladnoi Khimii 2000; 73: 714-18.
20. Pavlenko, S., Malyshkin, V., Bazhenov, Yu. *Cement-Free Small-Grain Concrete from Secondary Mineral Resources*. Novosibirsk: SB RAS, 2000.
21. Sulimenko L.M., Schalunenko N.J., Urkhanova L.A. Mechanochemical activation of binder compositions. Cement 1994; 2: 38-40.
22. Avvakumov E.G., Pavlenko S.I., Kosova N.V., et al. Compositional binder from mechanochemical activated industrial wastes. Chem. Sustainable Development 2000; 8: 636-40.
23. Kitamura M., Kamitani M., Senna M. Rapid hardening of cement by the addition of a mechanically activated $Al(OH)_3$ - $Ca(OH)_2$ mixure. J. Amer. Ceram. Soc. 2000; 83: 523-27.
24. Ilin A.P., Prokofiev V.Ju., Sazanova T.V., Kochetkov S.P. Creation of absorbers for adsorbtional purification of technological gases containing fluorine compounds. Zhurnal Prikladnoi Khimii 1999; 1489-92.

25. Sazanova T.V. Creation of sorbents for the purification of waste gases of industrial production from fluorine compounds. Cand. thesis. Ivanovo: Khimico-tekhnologicheskii Univ, 2000.
26. Kotsupalo N.P., Menzheres L.T., Mamilova E.V. Sorbtion properties of defect forms $LiCl{\cdot}2Al(OH)_3{\cdot}mH_2O$. Zhurnal Prikladnoi Khimii 1998; 72: 16-25.
27. Kotsupalo N.P., Menzheres L.T., Mamilova E.V., Ryabtsev A.D. Preparation method of sorbent $LiCl{\cdot}2Al(OH)_3{\cdot}mH_2O$ for lithium extraction from pickles. Chem. Sustainable Development 1999; 7: 249-53.
28. Patent RU 2050330. Method of lithium selective sorbtion exraction from pickles and mounting for realization. Ryabtsev A.D., Menzheres L.T., Kotsupalo N.P. et al.
29. PCT (WO) 94/19280. Preparation method of sorbent for lithium extraction from pickles. Kotsupalo N.P., Menzheres L.T., Sitnicova L.L.
30. Ryabtsev A.D., Menzheres L.T., Kotsupalo N.P. et al. Preparation of granular sorbent on base of $LiCl{\cdot}2Al(OH)_3{\cdot}mH_2O$ unwasted method. Chem. Sustainable Development 1999; 7: 343-48.
31. Kotsupalo N.P., Menzheres L.T., Ryabtsev A.D. Choice of complex technology for exploit of pickles of calcium chloride type. Chem. Sustainable Development 1999; 7: 157-62.
32. Kotsupalo N.P., Menzheres L.T., Ryabtsev A.D. Concept of complex utilisation of pickles of calcium chloride type. Chem. Sustainable Development 1999; 7: 57-61.
33. Patent USA 4812300. Selective perovskite catalysts to oxidize ammonia to nitric oxide. Quinian M.A., Ramanathan R., Wise H.
34. Patent RU 2100068. Catalyst of ammonia oxidation. Isupova L.A., Sadykov V.A., Kosova N.V., Avvakumov E.G. et al.
35. Pauli I.A., Avvakumov E.G., Poluboyarov V.A. et al. Influence of mechanical activation on the synthesis and catalytic properties of lanthanum cobaltite. Sib. Khim. Zhurn. 1992; 3: 133-37.
36. Kosova N.V., Devyatkina E.T., Avvakumov E.G. et al. Mechanochemical synthesis of calcium ferrite with perovskite structure. Neorg. Materialy 1998; 34: 478-81.
37. Isupova L.A., Sadykov V.A., Avvakumov E.G., Kosova N.V. Mechanical activation in technology of oxide catalysts. Chem. Sustainable Development 1998; 6: 207-10.

# Conclusions

The monograph is devoted to development of new nontraditional method of preparation of advanced oxide materials, exhibiting a broad range of exploitation characteristics and, thus, being promising for new technologies. Continuously growing interest for new materials is the driving force of the progress and improvement of technological processes of their synthesis, providing increase of the yield and the quality of products. The proposed method was called soft mechanochemical synthesis. It is based on the use of compounds reacting via acid-base mechanism under the mechanical treatment in high energetic grinding apparatus followed by calcination at moderate temperatures.

The monograph represents theoretical and experimental foundations of this method. One of the main characteristics of acid-base properties of compounds is electronegativity. According to Pauling, it is calculated from thermodynamic data; according to Mulliken, - from ionization potential and electron affinity ('orbital electonegativity'); and according to Godovikov, – from ionization potential, ion radius and coordination number ('energy parameter'). The authors of the present monograph show that a linear correlation for the difference between the sums of average orbital electronegativities of the final and initial products, and Gibbs energies of the reactions exists. On the basis of this dependence, Gibbs energies of the reactions can be estimated. It is especially important in case when thermodynamical data for initial and final products are absent.

A correlation was stated between the difference in acid-base properties estimated on the basis of electronegativity concept and the rate of mechanochemical interaction obtained experimentally.

In the monograph, the experimental data on soft mechanochemical synthesis of double and multicomponent oxides obtained by the present authors and other researchers have been systematized. The specific features and regularities of the reactions of soft mechanochemical synthesis have been established. It was demonstrated that at the first stage the interaction proceeds with the participation of surface acid and base centres, while at the second stage, the process goes on with the participation of dissolved state since water is formed in the result of interaction, i.e. according to hydrothermal-like mechanism.

The authors show that the adding of free water influences the mechanochemical interactions in the mixtures of anhydrous and hydrated oxides. The maximum interaction rate is observed at optimal (low, as a rule,) concentration of water, thus, revealing a new type of interactions of solids at the presence of small amounts of water differing from those in the mixtures of solids and in aqueous solutions.

It was demonstrated that the use of mechanical activation allows to decrease the temperature of consequent thermal treatment by 200–300°C and to decrease the annealing time, thus, providing the materials of high homogeneity and high dispersion. The quality of materials as prepared is similar or even better than of those prepared by traditional methods of synthesis. The perspectives of the method for chemical industry have been demonstrated using specific examples.

The monograph also deals with the description of high energetic laboratory and semi-industrial apparatus constructed and produced in Russia and Japan which are widely used for efficient performance of mechanochemical reactions.

The advanced method of soft mechanochemical synthesis as compared with the others is characterized by higher environmental purity, the absence of the waste solutions and sewage that pose major problems for their recovery. The method also provides the decrease of energy consumption since no prolonged activation and calcination at high temperatures are needed.

## Subject Index

# Author Index

GPSR Compliance

*The European Union's (EU) General Product Safety Regulation (GPSR) is a set of rules that requires consumer products to be safe and our obligations to ensure this.*

*If you have any concerns about our products, you can contact us on ProductSafety@springernature.com*

In case Publisher is established outside the EU, the EU authorized representative is:

Springer Nature Customer Service Center GmbH
Europaplatz 3
69115 Heidelberg, Germany

**Batch number: 10584448**

Printed by Printforce, the Netherlands